Springer Theses

Recognizing Outstanding Ph.D. Research

Aims and Scope

The series "Springer Theses" brings together a selection of the very best Ph.D. theses from around the world and across the physical sciences. Nominated and endorsed by two recognized specialists, each published volume has been selected for its scientific excellence and the high impact of its contents for the pertinent field of research. For greater accessibility to non-specialists, the published versions include an extended introduction, as well as a foreword by the student's supervisor explaining the special relevance of the work for the field. As a whole, the series will provide a valuable resource both for newcomers to the research fields described, and for other scientists seeking detailed background information on special questions. Finally, it provides an accredited documentation of the valuable contributions made by today's younger generation of scientists.

Theses are accepted into the series by invited nomination only and must fulfill all of the following criteria

- They must be written in good English.
- The topic should fall within the confines of Chemistry, Physics, Earth Sciences, Engineering and related interdisciplinary fields such as Materials, Nanoscience, Chemical Engineering, Complex Systems and Biophysics.
- The work reported in the thesis must represent a significant scientific advance.
- If the thesis includes previously published material, permission to reproduce this must be gained from the respective copyright holder.
- They must have been examined and passed during the 12 months prior to nomination.
- Each thesis should include a foreword by the supervisor outlining the significance of its content.
- The theses should have a clearly defined structure including an introduction accessible to scientists not expert in that particular field.

More information about this series at http://www.springer.com/series/8790

Suman Baral

Thomas-Fermi Model for Mesons and Noise Subtraction Techniques in Lattice QCD

Doctoral Thesis accepted by the Baylor University, Texas, USA

 Springer

Suman Baral
Everest Institute of Science and Technology
Kathmandu, Nepal

Neural Innovations LLC
Lorena, TX, USA

ISSN 2190-5053   ISSN 2190-5061   (electronic)
Springer Theses
ISBN 978-3-030-30906-0   ISBN 978-3-030-30904-6   (eBook)
https://doi.org/10.1007/978-3-030-30904-6

To my parents
  Dadhi Raj Baral and Geeta Baral,
  and my brother
  Sujan Baral
  for always believing in me...

Supervisor's Foreword

It is a pleasure to write this Springer Thesis foreword for Dr. Suman Baral, my former Ph.D. student originally from Nepal. In theoretical physics it is important for advanced students to be aware of both the mathematical/numerical and physical aspects of scientific research. Thus, I have required my doctoral students to contribute to both a mathematically oriented and a more traditional physics-related work in their Ph.D. dissertations. We are greatly aided in the mathematics part by our ongoing collaboration with Dr. Ronald Morgan of the Baylor University Mathematics Department.

The more mathematically related subject in this dissertation is in the area of lattice QCD (quantum chromodynamics) noise subtraction (or suppression) techniques. Lattice QCD is the numerical approach to solving the theory of strong interactions on a space-time lattice of points. One of the most difficult problems in lattice measurements of many physical properties is the issue of evaluating quark loop effects. These can be thought of as arising from Feynman-like diagrams which have so-called "disconnected" parts, which are really quark loops that are connected to other particles in the diagram by gluon lines. Such diagrams are the hardest to simulate on supercomputers, and special algorithms are required for their efficient evaluation. Using the numerical techniques developed by Dr. Morgan, we employ matrix "deflation" algorithms to reduce statistical uncertainty in these time-consuming lattice calculations. In an extension to Suman's work, we show that this allows us to accelerate the calculation of disconnected quantities by over an order of magnitude[1]. In addition, the technique works at the lowest possible quark masses. This work is contained in the first part of Suman's dissertation.

In the more physics-related work of this dissertation, Suman and I have developed a new model for mesonic matter using the Thomas-Fermi (TF) statistical quark model. I and another student, Dr. Quan Liu, previously developed a baryonic matter model with this same approach. A major motivation of this study is the "tetraquark" states recently discovered by various experimental phenomenology groups and the

---

[1] S. Baral, T. Whyte, W. Wilcox, R. Morgan, Comput. Phys. Commun. **241**, 64 (2019)

possibility that additional stable multi-quark families exist. Similar to our previous TF baryonic study, we use a two-inequivalent wave function approach to investigate aspects of multiquark matter. We think of our model as a tool for quickly assessing the characteristics of new, possibly bound, particle states of higher quark number content, which cannot yet be examined by lattice methods. Using parameters fit from heavy-light quark systems, we have observed interesting patterns of single-quark energies, showing the possibility that a family stability of such particles exists. Our methods can also be extended to combinations of five quark states called "pentaquarks."

It has been a pleasure to work with Suman on these subjects. This Springer publication serves to recognize his contributions and the promise of these approaches.

Professor of Physics   Walter Wilcox
Baylor University, Waco, TX, USA
Secretary-Treasurer, Texas Section American Physical Society

Contents

Chapter 1
New Noise Subtraction Methods for Lattice QCD Calculations

1.1 Introduction

Lattice QCD is a set of numerical techniques which uses a finite space-time lattice to simulate the interactions between quarks and gluons. In lattice QCD, one of the important and challenging issues is calculation of disconnected quark loops. In order to calculate them many matrix inversions are required. To tackle this problem, instead of a direct approach, we use isolated stochastic approach along with approximation techniques [1]. Noise theory is used to project out the loop operator expectation values and new noise subtraction methods are applied to reduce the statistical uncertainty. Perturbative subtraction [2] is a standard noise subtraction method which we will be comparing to and attempting to improve upon. This paper will be focusing on eigenspectrum subtraction (deflation), polynomial subtraction, and combination methods.

1.2 Noise Methods

Information about a system can be extracted by projecting its signal on the randomly generated noise vectors. For example, a QCD matrix M can be projected over randomly chosen noise vector η to extract the solution vector x as we can see in the equation

$$Mx = \eta, \tag{1.1}$$

where M is of size $\mathcal{N} \times \mathcal{N}$, x and η are of size $\mathcal{N} \times 1$.

If α and β are two arbitrary vectors of size $\mathcal{N} \times \mathcal{L}$ in space-time-color-Dirac space, then (ij)th element of the product $\langle \alpha\beta \rangle$ can be written as

© Springer Nature Switzerland AG 2019
S. Baral, *Thomas-Fermi Model for Mesons and Noise Subtraction Techniques in Lattice QCD*, Springer Theses, https://doi.org/10.1007/978-3-030-30904-6_1

$$\langle \alpha_i \beta_j \rangle \equiv \lim_{N \to \infty} \frac{1}{N} \sum_n^N \alpha_i^{(n)} \beta_j^{*(n)}. \tag{1.2}$$

Here α_i and β_j are the row matrices of size $1 \times \mathcal{L}$, $n = 1, 2, \ldots, N$ represents N set of noise vectors over which arbitrary vectors are projected. It is divided by N to average over all the noises used. When $N \to \infty$ we get the exact result; however, due to computational limitations, N has to be chosen to be finite and the solution we obtained is approximate rather than exact. In numerical calculations we use Eq. (1.3) instead of Eq. (1.2) where

$$\langle \alpha_i \beta_j \rangle \approx \frac{1}{N} \sum_n^N \alpha_i^{(n)} \beta_j^{*(n)}. \tag{1.3}$$

For a sufficiently large value of N, if we apply Eq. (1.2) to noise vectors, we will get

$$\langle \eta_i \eta_j \rangle = \lim_{N \to \infty} \frac{1}{N} \sum_n^N \eta_i^{(n)} \eta_j^{*(n)} = \delta_{ij}. \tag{1.4}$$

Also, expectation value of the noise vectors yields

$$\langle \eta_i \rangle = \lim_{N \to \infty} \frac{1}{N} \sum_n^N \eta_i^{(n)} = 0. \tag{1.5}$$

These two properties of noise, along with indexed form of Eqs. (1.1) and (1.2) help us determine each element from the quark matrix inverse, M^{-1}, which we will see from the equation below.

$$\begin{aligned}
M_{ik}^{-1} &= \sum_j M_{ij}^{-1} \delta_{jk} \\
&= \sum_j M_{ij}^{-1} \langle \eta_j \eta_k \rangle \\
&= \sum_j M_{ij}^{-1} \left(\lim_{N \to \infty} \frac{1}{N} \sum_n^N \eta_j^{(n)} \eta_k^{*(n)} \right) \\
&= \lim_{N \to \infty} \frac{1}{N} \sum_n^N \sum_j M_{ij}^{-1} \eta_j^{(n)} \eta_k^{*(n)} \\
&= \lim_{N \to \infty} \frac{1}{N} \sum_n^N \left(\sum_j M_{ij}^{-1} \eta_j^{(n)} \right) \eta_k^{*(n)}
\end{aligned} \tag{1.6}$$

$$= \lim_{N \to \infty} \frac{1}{N} \sum_n^N x_i^{(n)} \eta_k^{*(n)}$$

$$= \langle x_i \eta_k \rangle.$$

As N is not equal to infinity in our computer simulations, we can see that inverse of the matrix calculated is not exact and hence there is an error associated with it. We apply several techniques that attempt to minimize statistical uncertainty which will be discussed further in this dissertation.

We have used $Z(N)$ noises, with $N = 4$, for our computer simulations. Each of the N values in $Z(N)$ noise is separated by $360/N$ degrees. For $N = 4$, the separation is 90°. We settled with 0°, 90°, 180°, and 270° even though we could choose any four points separated by 90°. In the complex plane, using the relation

$$x + iy = re^{i\theta} \tag{1.7}$$

these points become $1, i, -1$, and $-i$, respectively, for the unit radius r. This means all the random noises that we generated had only four distinct elements calculated above.

1.3 Variance

Many operators in lattice QCD simulations are flooded with noise. Several strategies could be applied to reduce the variance of these operators, which originates in the off-diagonal components of the associated quark matrix. One basic strategy is to mimic the off-diagonal elements of the inverse of the quark matrix with another traceless matrix, thereby maintaining the trace but with reduced statistical uncertainty. In this section, we will explain the theory behind it.

If α and β from Eq. (1.3) are replaced with two different noise vectors, we get

$$X_{ij} = \langle \eta_i \eta_j \rangle \approx \frac{1}{N} \sum_n^N \eta_i^{(n)} \eta_j^{*(n)} = \delta_{ij}, \tag{1.8}$$

where X from now on represents the matrix made from projection of two noise vectors.

$$X_{ij} = \frac{1}{N} \sum_{n=1}^N \eta_{in} \eta_{jn}^* \tag{1.9}$$

It can be shown that

$$V[Tr(QX)] \equiv \left\langle \left| \sum_{i}^{N} q_{ij} X_{ji} - Tr(Q) \right|^2 \right\rangle$$

$$= \sum_{i \neq j} \left(\left\langle |X_{ji}|^2 \right\rangle |q_{ij}|^2 + q_{ij} q_{ji}^* \left\langle (X_{ji})^2 \right\rangle \right) + \sum_{i} \left\langle |X_{ii} - 1|^2 \right\rangle |q_{ii}|^2 ,$$

(1.10)

where Q is the matrix-representation of an operator. First, let us consider a general real noise. The constraints are

$$\left\langle |X_{ji}|^2 \right\rangle = \left\langle (X_{ji})^2 \right\rangle = \frac{1}{N}$$

(1.11)

for $i \neq j$.

For general real noises variance can now be written as

$$V[Tr(QX_{real})] = \frac{1}{N} \sum_{i \neq j} \left(|q_{ji}|^2 + q_{ij} q_{ji}^* \right) + \sum_{i} \left\langle |X_{ii} - 1|^2 \right\rangle |q_{ii}|^2 .$$

(1.12)

For $Z(2)$ noises apart from the constraints at Eq. (1.11), there is one additional constraint, $\langle |X_{ii} - 1|^2 \rangle = 0$ for $i = j$. This gives

$$V\left[Tr\left(QX_{z(2)}\right)\right] = \frac{1}{N} \sum_{i \neq j} \left(|q_{ji}|^2 + q_{ij} q_{ji}^* \right).$$

(1.13)

For $Z(4)$ noises, the constraints are little different. That is, the constraints

$$\left\langle |X_{ji}|^2 \right\rangle = \frac{1}{N},$$

$$\left\langle (X_{ji})^2 \right\rangle = 0,$$

$$\left\langle |X_{ii} - 1|^2 \right\rangle = 0$$

(1.14)

make the variance at Eq. (1.10) to be

$$V\left[Tr\left(QX_{z(N \geq 3)}\right)\right] = \frac{1}{N} \sum_{i \neq j} |q_{ji}|^2 .$$

(1.15)

From this equation we can see variance only depends on off-diagonal elements of the matrix for $Z(4)$ noises. This relation is the foundation for subtraction noise techniques that we develop.

1.4   Subtraction Noise Techniques

Noise subtraction techniques can be used to reduce the error associated with disconnected operators. In lattice QCD this is equivalent to reducing the variance of trace of inverse of the quark matrix. These methods reduce the computational needs as well because in these methods less matrix inversions are required to achieve the desired variance.

Given two matrices Q and $\tilde{Q}$, where $\tilde{Q}$ is traceless, we can show that expectation value of the trace of a matrix remains unchanged after adding a traceless matrices to it. That is, the relationship

$$\langle Tr(QX) \rangle = \left\langle Tr\left\{ \left(Q - \tilde{Q} \right) X \right\} \right\rangle \tag{1.16}$$

holds true for large number of noise. Note that average is taken only over noises and using the identity matrix, $\langle X_{ij} \rangle = \delta_{ij}$, Eq. (1.16) can be easily established.

We just observed that trace remains invariant under addition of a traceless matrix. If we take the variance of the trace, though, the relationship does not hold true. We can see from Eq. (1.15) that variance depends on the off-diagonal elements of a matrix. So, if we can find a matrix $\tilde{Q}$ which is traceless but has the off-diagonal elements same as that of Q, new matrix $Q - \tilde{Q}$ is a diagonal traceless matrix which would have zero variance. This can be explained from the relation

$$V\left[Tr\left\{ (Q - \tilde{Q}) X_{z(N \geq 3)} \right\} \right] = \frac{1}{N} \sum_{i \neq j} \left(|q_{ij} - \tilde{q}_{ij}|^2 \right). \tag{1.17}$$

If q_{ij} and $\tilde{q}_{ij}$ are both equal for $i \neq j$, then Eq. (1.17) gives a zero variance. However, this situation is ideal. In practical conditions, it is much difficult to mimic the off-diagonal elements of the quark matrix.

In case of lattice QCD calculations, Q is actually the inverse of the quark matrix M. So, we have to find a matrix $\tilde{M}^{-1}$ that is traceless and has off-diagonal elements as close to M^{-1} as possible. We use six different techniques to create $\tilde{M}^{-1}$ which will eventually determine the method that works best on reducing the variance. These methods will be described in the following section.

1.5   Methods

We have applied several new techniques to non-subtracted (NS) lattice data. These are termed eigenvalue subtraction (ES) [3], Hermitian forced eigenvalue subtraction (HFES) [4–6], and polynomial subtraction (POLY) [7]. In [4] we also introduced techniques which combine deflation with other subtraction methods. Here we will compare various methods and show how effective the combination methods are.

We work with the standard Wilson matrix in the quenched approximation. The size of the lattice first worked on is 8^4 and after looking at the trends we switch to a bigger one of size $24^3 \times 32$. We use 200 $Z4$ noises for three different set of kappa values 0.155, 0.156, and 0.157. Linear equations are solved using GMRES-DR (generalized minimum residual algorithm-deflated and restarted) for the first noise and GMRES-Proj (similar algorithm projected over eigenvectors) for remaining noises [8]. All of our methods attempt to design a traceless matrix $\tilde{M}^{-1}$ in order to obtain off-diagonal elements as close to M^{-1} as possible.

Expectation value of any operator can be written as

$$\langle \bar{\psi} \mathcal{O} \psi \rangle = -Tr(\mathcal{O} M^{-1}). \tag{1.18}$$

For the scalar operator calculations, $\mathcal{O}$ can be replaced by identity operator and expectation value can be written as

$$\langle \bar{\psi}(x) \psi(x) \rangle = -Tr(M^{-1}). \tag{1.19}$$

Trace is defined as

$$Tr(M^{-1}) \equiv \sum_i M_{ii}^{-1}. \tag{1.20}$$

Using Eq. (1.6), trace of the inverse can be expressed as

$$
\begin{aligned}
Tr(M^{-1}) &= \sum_i \langle x_i \eta_i \rangle \\
&= \sum_i \frac{1}{N} \sum_n^N x_i^{(n)} \eta_i^{*(n)} \\
&= \sum_i \frac{1}{N} \sum_n^N \left(\sum_j M_{ij}^{-1} \eta_j^{(n)} \right) \eta_i^{*(n)} \\
&= \sum_i \frac{1}{N} \sum_n^N \eta_i^{*(n)} \sum_j M_{ij}^{-1} \eta_j^{(n)}.
\end{aligned}
\tag{1.21}
$$

Calculation of trace using this equation is equivalent to vanilla noise method, also called non-subtraction method (NS). All of our methods will try to better this method. This is done by creating a traceless matrix $\tilde{M}^{-1}$ that is pretty close to M^{-1} especially on the off-diagonal elements. Closer it is to the matrix, better the method turns out to be. Addition of a traceless matrix $\tilde{M}^{-1}$ will now modify the relationship as

$$Tr(M^{-1}) = \sum_i \frac{1}{N} \sum_n^N \eta_i^{*(n)} \sum_j \left(M_{ij}^{-1} - \tilde{M}_{ij}^{-1} \right) \eta_j^{(n)}. \tag{1.22}$$

In real world simulations, it is almost impossible to generate traceless matrices $\tilde{M}^{-1}$. So, for the correct evaluation of the expectation value of the operators $Tr(\tilde{M}^{-1})$ needs to be added to the calculation. This gives

$$Tr(M^{-1}) = \sum_i \frac{1}{N} \sum_n^N \eta_i^{*(n)} \sum_j \left(M_{ij}^{-1} - \tilde{M}_{ij}^{-1} \right) \eta_j^{(n)} + Tr \left(\tilde{M}^{-1} \right). \tag{1.23}$$

Simplification of this equation yields

$$Tr(M^{-1}) = \frac{1}{N} \sum_n^N \left(\eta^{(n)\dagger} \left(x^{(n)} - \tilde{x}^{(n)} \right) \right) + Tr \left(\tilde{M}^{-1} \right), \tag{1.24}$$

where $x^{(n)}$ is the solution vector generated when implementing the GMRES algorithms and $\tilde{x}^{(n)}$ is given by

$$\tilde{x}^{(n)} \equiv \tilde{M}^{-1} \eta^{(n)}. \tag{1.25}$$

For any operator Θ, the appropriate trace becomes

$$Tr \left(\Theta M^{-1} \right) = \frac{1}{N} \sum_n^N \left(\eta^{(n)\dagger} \Theta \left(x^{(n)} - \tilde{x}^{(n)} \right) \right) + Tr \left(\Theta \tilde{M}^{-1} \right). \tag{1.26}$$

Note that adding $Tr \left(\Theta \tilde{M}^{-1} \right)$ has no influence on the noise error bar, which is what is studied here. We will now explain different ways of obtaining the subtraction matrix $\tilde{M}^{-1}$.

1.5.1 Perturbative Subtraction (PS)

Perturbative noise subtraction method is the standard method used by researchers to estimate off-diagonal elements of the quark matrix. In this subsection we will explain how it does the estimation and in what regard it differs between local and nonlocal vectors.

For Wilson case, the quark matrix M, can be written as

$$M_{ij} = \delta_{ij} - \kappa P_{ij}, \tag{1.27}$$

where i, j covers space-time-color-Dirac space and P_{ij} is

$$P_{ij} = \sum_{\mu} \left[(1 - \gamma_{\mu}) U_{\mu}(x) \delta_{x,y-a_{\mu}} + (1 + \gamma_{\mu}) U_{\mu}^{\dagger}(x - a_{\mu}) \delta_{x,y+a_{\mu}} \right]. \quad (1.28)$$

If we take inverse of Eq. (1.27), we get

$$M_{ij}^{-1} = \frac{1}{\delta_{ij} - \kappa P_{ij}}. \quad (1.29)$$

We now can expand this equation in Taylor series for small values of κP to get the matrix $\tilde{M}_{pert}^{-1}$ such that

$$\tilde{M}_{pert}^{-1} = I + \kappa P + (\kappa P)^2 + (\kappa P)^3 + (\kappa P)^4 + \cdots \quad (1.30)$$

For simulations, we cannot calculate all the orders of $\tilde{M}_{pert}^{-1}$ and will have to settle for a finite order. Settling for the finite order will deny us from the matrix that is completely off-diagonal though. So, $Tr\left(\tilde{M}_{pert}^{-1}\right)$ will have to be added back for the correct trace computation. In this dissertation, the expansion is calculated up to seventh order of magnitude. That is,

$$\tilde{M}_{pert}^{-1} = I + \kappa P + (\kappa P)^2 + (\kappa P)^3 + (\kappa P)^4 + (\kappa P)^5 + (\kappa P)^6 + (\kappa P)^7 \quad (1.31)$$

As we can see in the equation,

$$\langle \bar{\psi}(x)\psi(x) \rangle = -Tr\left(M^{-1}\right), \quad (1.32)$$

only closed loop gauge invariant objects contribute to the trace. This means only closed path objects with an area A contribute to the trace in Eq. (1.33) which gives,

$$Tr\left(M^{-1}\right) = \frac{1}{N} \sum_{n}^{N} \left(\eta^{(n)\dagger} \cdot \left(x^{(n)} - \tilde{x}_{pert}^{(n)}\right)\right) + Tr\left(\tilde{M}_{pert}^{-1}\right), \quad (1.33)$$

Generalizing for any operator Θ, trace now takes the form,

$$Tr\left(\Theta M^{-1}\right) = \frac{1}{N} \sum_{n}^{N} \left(\eta^{(n)\dagger} \cdot \Theta \left(x^{(n)} - \tilde{x}_{pert}^{(n)}\right)\right) + Tr\left(\Theta \tilde{M}_{pert}^{-1}\right), \quad (1.34)$$

where

$$\tilde{x}_{pert}^{(n)} \equiv \tilde{M}_{pert}^{-1} \eta^{(n)}. \quad (1.35)$$

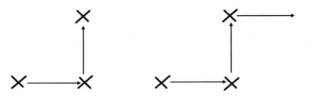

Fig. 1.1 Perturbative expansion contributions of $0(\kappa^2)$ and $0(\kappa^3)$

Fig. 1.2 Perturbative local operator contribution at $0(\kappa^4)$ and $0(\kappa^6)$

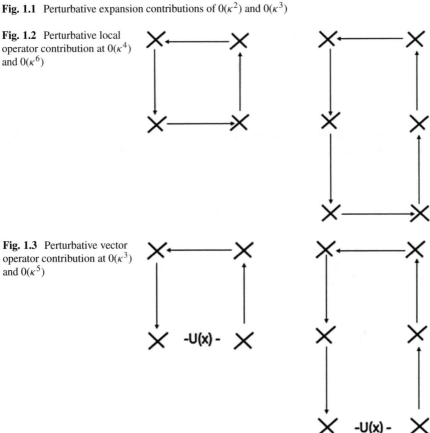

Fig. 1.3 Perturbative vector operator contribution at $0(\kappa^3)$ and $0(\kappa^5)$

Geometric interpretation of perturbative expansion in Fig. 1.1 shows how each order of κ is related to a link.

We can see from Fig. 1.2 that local operators require a correction starting at 4th order of κ and from Fig. 1.3 that nonlocal operators require a correction starting at 3rd order of κ because the least number of links required to form a closed loop in local operator and nonlocal operator is 4 and 3, respectively. Note that point-split operator has already an inbuilt order of κ. Also, local scalar operator requires a correction starting at 0th order of κ. In general, even orders of κ contribute to the local operators and odd orders contribute to the nonlocal operators.

The direct calculation of the $Tr\left(\Theta M_{pert}^{-1}\right)$ is very expensive; however, closed loops can be easily calculated. This can be done by solving the system,

$$M_{pert}x = e_i, \tag{1.36}$$

that does not involve a noise vector, where e_i is the unit vector in the ith direction that spans space-time-color-Dirac space. Now, we can solve for the solution vector by calculating

$$x = \tilde{M}_{pert}^{-1}e_i = \left(I + \kappa P + (\kappa P)^2 + (\kappa P)^3 + \cdots + (\kappa P)^n\right)e_i. \tag{1.37}$$

Not all the $O(\kappa^n)$ terms contribute to the calculation of trace for each specific operator. We will then drop the terms that do not contribute to trace. That means for the calculation of nonlocal operators we drop the even orders of κ, while odd orders of κ are dropped for calculation of local scalar and local vector operators. We denote the truncated expansion as $\hat{M}_{pert}^{-1}$. Trace of the corrected part is now calculated as

$$Tr\left(\Theta M_{pert}^{-1}\right) = \sum_i \left(e_i^*\Theta\hat{M}_{pert}^{-1}e_i\right). \tag{1.38}$$

We worked up to sixth order of κ in local and fifth order in nonlocal operators. Higher order subtractions can also be considered. The effect of higher order subtractions has been studied previously by our group [7].

1.5.2 Polynomial Subtraction (POLY)

Perturbative noise subtraction method is widely used not only because it reduces the error associated with lattice QCD simulations but also because $\tilde{M}_{pert}^{-1}$ is relatively easier to build. This fact motivated our group to develop a new technique called Polynomial Noise Subtraction method, also referred as POLY in this dissertation, which is very similar to PS method. The only difference is that coefficients of the new matrix that mimics the off-diagonal elements are now different from one. That is, $\tilde{M}_{pert}^{-1}$ which was initially represented to the 6th order as

$$\tilde{M}_{pert}^{-1} \equiv 1 + \kappa P + (\kappa P)^2 + (\kappa P)^3 + (\kappa P)^4 + (\kappa P)^5 + (\kappa P)^6 \tag{1.39}$$

is now replaced by $\tilde{M}_{poly}^{-1}$ such that

$$\tilde{M}_{poly}^{-1} \equiv a_1 + a_2\kappa P + a_3(\kappa P)^2 + a_4(\kappa P)^3 + a_5(\kappa P)^4 + a_6(\kappa P)^5 + a_7(\kappa P)^6, \tag{1.40}$$

where the a_i's are the coefficients obtained using MINimal RESidual (MinRes) polynomial [7]. Let us explain how we can obtain these coefficients.

The linear equation system we discuss in this dissertation that relates quark matrix, noise vector, and solution vector is

$$Mx = \eta. \tag{1.41}$$

Now we would want to create a Krylov subspace $\mathcal{K}_t$ spanned by b, Mb, M^2b, M^3b, ... such that

$$\mathcal{K}_t = \{b, Mb, M^2b, M^3b, \ldots\}. \tag{1.42}$$

Any solution vector x_t can also be expressed as a linear combination of elements from this subspace. That is,

$$\begin{aligned} x_t &= a_0 b + a_1 M b + a_2 M^2 b + a_3 M^3 b + \cdots \\ &= \left(a_0 + a_1 M + a_2 M^2 + a_3 M^3 + \cdots \right) b \\ &= P(M) b, \end{aligned} \tag{1.43}$$

where $P(M)$ is a polynomial of M.

We want to minimize norm of the residual $||Mx_t - b||_2$ which now equals $||(MP(M) - I)b||_2$. We can see that after residual is minimized, we get that the polynomial is equivalent to the inverse of the quark matrix. This can be understood from the equation below.

$$\begin{aligned} ||Mx_t - b||_2 &\approx 0 \\ \implies ||(MP(M) - I)b||_2 &\approx 0 \\ \implies P(M) &\approx M^{-1} \end{aligned} \tag{1.44}$$

In POLY method, we use this polynomial $P(M)$ obtained from the MinRes projection to mimic off-diagonal elements of the inverse of a quark matrix. As will see this method produces smaller variance than the PS method without significant increment of matrix vector products.

Consider nth order polynomial of M such that

$$P_n(M) = a_0 + a_1 M + a_2 M^2 + \cdots + a_n M^n, \tag{1.45}$$

where the coefficients

$$\vec{a} = \{a_0, a_1, \ldots, a_n\} \tag{1.46}$$

are obtained from MinRes projection.

By solving a set of equation

$$\left(v^\dagger v\right)\vec{a} = v^\dagger b,\tag{1.47}$$

where v is a matrix given by

$$v = \begin{pmatrix} Mb & M^2 b & \dots & M^{n+1} b \end{pmatrix},\tag{1.48}$$

we can obtain the coefficients and build a polynomial $P(M)$. From Eq. (1.44) this is approximately equal to inverse of a quark matrix, M^{-1}. So this polynomial is used to construct $\tilde{M}^{-1}_{poly}$ which is used exactly as $\tilde{M}^{-1}_{pert}$ does in the perturbative subtraction method. Coefficients only need to be calculated for one single noise vector and then used for all the right-hand sides. The shape of 7th order MinRes polynomial for 8^4 lattice and $\kappa = 0.157$ is shown in Fig. 1.4 and the coefficients are listed in Table 1.1. Random noise vector was used as the right-hand side of the system. From the table we can see sign of the consecutive coefficients is alternating between positive and negative. Also, these coefficients have a larger value for middle order coefficients and far different from one as seen in PS method. We also noticed that coefficients did not change significantly from configuration to configuration. We can see from Fig. 1.4 that polynomial is a good approximation to M^{-1} since it remains flat at $z = 1$ plane in some region of the complex space.

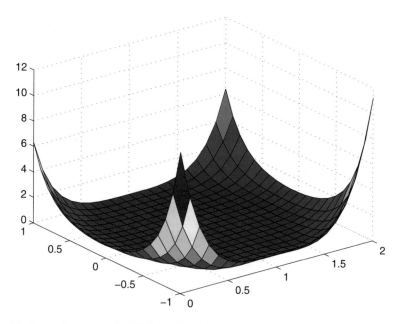

Fig. 1.4 Shape of seventh order MinRes polynomial

Table 1.1 Coefficients of the
MinRes polynomial

Coefficients	Real part	Imaginary part
a_0	6.4551	0.0263
a_1	−20.5896	−0.1429
a_2	39.3584	0.3563
a_3	−47.6275	−0.5076
a_4	36.7205	0.4378
a_5	−17.4702	−0.2265
a_6	4.6727	0.0648
a_7	−0.5375	−0.0079

Since $M = I + \kappa P$, we can express $\tilde{M}_{poly}^{-1}$ in terms of P as shown below

$$
\begin{aligned}
\tilde{M}_{poly}^{-1} &= a_0 + a_1 M + a_2 M^2 + \cdots + a_n M^n \\
&= a_0 + a_1 (I + \kappa P) + a_2 (I + \kappa P)^2 + \cdots + a_n (I + \kappa P)^n \quad (1.49) \\
&= b_0 + b_1 \kappa P + b_2 \kappa^2 P^2 + \cdots + b_n \kappa^n P^n.
\end{aligned}
$$

These coefficients $\{b_0, b_1, \ldots, b_n\}$ can easily be calculated from the coefficients $\{a_0, a_1, \ldots, a_n\}$. As we can see the pattern for $\tilde{M}_{poly}^{-1}$ is quite similar to that of $\tilde{M}_{pert}^{-1}$, difference only being that coefficients in the latter are equal to one, we can build truncated M_{poly}^{-1} in the same way as we did for $\tilde{M}_{pert}^{-1}$ by dropping the terms that do not contribute to trace calculation.

Then the correction term is

$$
Tr\left(OM_{poly}^{-1}\right) = \sum_i \left(e_i^* O \tilde{M}_{poly}^{-1} e_i\right). \quad (1.50)
$$

The trace in this method takes the form

$$
Tr\left(\Theta M^{-1}\right) = \frac{1}{N} \sum_n^N \left(\eta^{(n)\dagger} \cdot \Theta \left(x^{(n)} - \tilde{x}_{poly}^{(n)}\right)\right) + Tr\left(\Theta \tilde{M}_{poly}^{-1}\right), \quad (1.51)
$$

where

$$
\tilde{x}_{poly}^{(n)} \equiv \tilde{M}_{poly}^{-1} \eta^{(n)}. \quad (1.52)
$$

1.5.3   Eigenvalue Subtraction (ES)

Perturbative noise subtraction and polynomial subtraction both involve an expansion of κ which means only small values of κ can be effectively used. These small values of κ lead to large, non-physical, quark masses. If we want to work for a higher order of κ that corresponds to realistic quark masses, we need to have a subtraction method that works best in this region. Eigenvalue subtraction is one such method developed by our group to cope with this problem. We will see that this technique will get flooded by highly non-normal condition and does not work as expected, however, paves a path to develop a new method called Hermitian forced eigenvalue subtraction (HFES) method.

The spectrum of low eigenvalues of matrices can limit the performance of iterative solvers. We have emphasized the role of deflation in accelerating the convergence of algorithms in Ref. [10]. Here we investigate deflation effects in statistical error reduction. Consider the vectors $e_R^{(q)}$ and $e_L^{(q)\dagger}$, which are defined as normalized right and left eigenvectors of the matrix M, as in

$$M e_R^{(q)} = \lambda^{(q)} e_R^{(q)}, \tag{1.53}$$

and

$$e_L^{(q)\dagger} M = \lambda^{(q)} e_L^{(q)\dagger}, \tag{1.54}$$

where $\lambda^{(q)}$ is the eigenvalue associated with both eigenvectors. With a full set N of eigenvectors and eigenvalues the matrix M can be fully formed as

$$M = \sum_{q=1}^{N} e_R^{(q)} \lambda^{(q)} e_L^{(q)\dagger}, \tag{1.55}$$

or

$$M = V_R \Lambda V_L^\dagger, \tag{1.56}$$

where V_R contains the right eigenvectors and $V_L^\dagger$ contains the left eigenvectors. Λ is a purely diagonal matrix made up of the eigenvalues of M in the order they appear in both V_R and V_L.

ES method involves forming a matrix, $\tilde{M}^{-1}$ by utilizing eigenspectrum decomposition. It is computationally unfeasible to use all the eigenmodes of M which means we can form only an approximation to M^{-1} by using only the Q smallest pairs where Q is determined based on the computational resources. We name the subtraction matrix thus formed as $\tilde{M}_{eig}^{-1}$.

Deflating out eigenvalues with the linear equation solver GMRES-DR can mimic the low eigenvalue structure of the inverse of matrix M as

$$\tilde{M}_{eig}^{-1} \equiv \tilde{V}_R \tilde{\Lambda}^{-1} \tilde{V}_L^\dagger, \tag{1.57}$$

where $\tilde{V}_R$ and $\tilde{V}_L^\dagger$ are the computed right and left eigenvectors and $\tilde{\Lambda}^{-1}$ is the inverse of eigenvalues. Here, $\tilde{V}_R$ is $\mathcal{N} \times Q$, $\tilde{\Lambda}^{-1}$ is of size $Q \times Q$, and $\tilde{V}_L^\dagger$ is of size $Q \times \mathcal{N}$ which makes $\tilde{M}_{eig}^{-1}$ same size as M^{-1}.

In index form, Eq. (1.57) can be written as

$$\left[\tilde{M}_{eig}^{-1}\right]_{ij} = \sum_q^Q \left(\left(\sum_k^Q \tilde{V}_{Rik} \tilde{\Lambda}_{kq}^{-1} \right) \left(\tilde{V}_{Ljq} \right)^\dagger \right). \tag{1.58}$$

We can further simplify it using the relation $\tilde{\Lambda}_{kq}^{-1} = \lambda_k^{-1} \delta_{kq}$ such that

$$\begin{aligned}
\left[\tilde{M}_{eig}^{-1}\right]_{ij} &= \sum_q^Q \left(\left(\sum_k^Q \tilde{V}_{Rik} \lambda_k^{-1} \delta_{kq} \right) \left(\tilde{V}_{Ljq} \right)^\dagger \right) \\
&= \sum_q^Q \tilde{V}_{Riq} \lambda_q^{-1} \left(\tilde{V}_{Ljq} \right)^\dagger.
\end{aligned} \tag{1.59}$$

We choose smallest Q eigenvalues of the matrix M to form $\tilde{M}_{eig}^{-1}$ because their contributions are greatest to the trace. Application of this eigenspectrum formulation enables us to represent the off-diagonal elements for smaller quark masses also which was previously not possible under PS and POLY methods. This means the formulation should work well for large κ values also.

The vacuum expectation term, $Tr\left(\tilde{M}_{eig}^{-1}\right)$ cannot be calculated by analyzing the closed loops as we did in PS methods. Instead, they have to be calculated directly as

$$Tr\left(\tilde{M}_{eig}^{-1}\right) = \sum_q^Q \frac{1}{\lambda^{(q)}}. \tag{1.60}$$

The trace of the inverse matrix now takes the form

$$Tr\left(M^{-1}\right) = \frac{1}{N} \sum_n^N \left(\eta^{(n)\dagger} \left(x^{(n)} - \tilde{x}_{eig}^{(n)} \right) \right) + Tr\left(\tilde{M}_{eig}^{-1}\right), \tag{1.61}$$

and

$$Tr\left(\Theta M^{-1}\right) = \frac{1}{N} \sum_n^N \left(\eta^{(n)\dagger} \Theta \left(x^{(n)} - \tilde{x}_{eig}^{(n)} \right) \right) + Tr\left(\Theta \tilde{M}_{eig}^{-1}\right), \tag{1.62}$$

where

$$\tilde{x}_{eig}^{(n)} = \tilde{M}_{eig}^{-1}\eta^{(n)}$$

$$= \sum_q^Q \frac{1}{\lambda_q} e_R^{(q)} \left(e_L^{(q)}.\eta^{(n)}\right). \qquad (1.63)$$

This last operation does not add matrix vector products. The generation of eigenmodes only requires the superconvergence solution of a single right-hand side with GMRES-DR. As pointed out previously [3], there is a relation between the even–odd eigenvectors for the reduced system and the full eigenvectors. Other right-hand sides are accelerated with GMRES-Proj using the eigenvalues generated.

In the results we will see that if we näively subtract eigenvalues from a non-Hermitian matrix, we often end up expanding the size of error bars. This happens because many of the right-handed eigenvectors of a non-Hermitian matrix can point in the same direction, a condition referred to as "highly non-normal."

1.5.4 Hermitian Forced Eigenvalue Subtraction (HFES)

To avoid the "highly non-normal" problem we force our matrix to be formulated in a Hermitian manner. The easiest way for us to do this with the Wilson matrix is to multiply by the Dirac γ_5 matrix. We can then form the low eigenvalue structure of $M\gamma_5$ from these eigenvalues. We define

$$M' \equiv M\gamma_5. \qquad (1.64)$$

Inverse of this matrix is written as

$$M'^{-1} \equiv \gamma_5 M^{-1}. \qquad (1.65)$$

Original trace of the inverse matrix can be obtained from new matrix M' without biasing the overall answer. From the relation,

$$Tr(\gamma_5 M'^{-1}) = Tr(\gamma_5\gamma_5 M^{-1})$$

$$= Tr(M^{-1}), \qquad (1.66)$$

we can see Hermitian matrix M' can be used to obtain the trace values associated with M^{-1}. It is important for the algorithm to do the multiplication on the *right* such that $M' = M\gamma_5$, but not $M' = \gamma_5 M$ to avoid using cyclic properties which fail in finite noise space.

We can now form normalized eigenvectors $e'^{(n)}_R$, eigenvalues $\lambda'^{(n)}$, and solution vectors $\tilde{x}'^{(n)}_{eig}$ for this new Hermitian matrix and perform a calculation similar to the ES method, accounting properly for the extra γ_5 factors. The trace of any operator Θ for the HFES method takes the following form:

$$Tr\left(\Theta M^{-1}\right) = \frac{1}{N}\sum_n^N \left(\eta^{(n)\dagger}\Theta\left(x^{(n)} - \tilde{x}'^{(n)}_{eig}\right)\right) + Tr\left(\Theta\gamma_5\tilde{M}'^{-1}_{eig}\right), \quad (1.67)$$

where

$$\tilde{x}'^{(n)}_{eig} \equiv \gamma_5\tilde{M}'^{-1}_{eig}\eta^{(n)} = \gamma_5\sum_q^Q \frac{1}{\lambda'^{(q)}}e'^{(q)}_R\left(e'^{(q)\dagger}_R\eta^{(n)}\right) \quad (1.68)$$

and

$$\tilde{M}'^{-1}_{eig} \equiv \tilde{V}'_R\tilde{\Lambda}'^{-1}\tilde{V}'^{\dagger}_R. \quad (1.69)$$

$\tilde{V}'_R$ is a matrix whose columns are the Q smallest right eigenvectors of M'. $\tilde{\Lambda}'^{-1}$ is the diagonal matrix of size Q that contains the inverse of eigenvalues $1/\lambda'^{(q)}$ as the diagonal elements. The price paid here, similar to the ES method, is a single extra superconvergence on one right-hand side for the non-reduced Hermitian system M' with GMRES-DR to extract eigenvectors and eigenvalues.

1.5.5   Combination Methods (HFPOLY and HFPS)

We have developed two methods which combine the error reduction techniques of HFES with POLY and PS, called HFPOLY and HFPS. Näively, for POLY we could think of this method as a subtracted combination: $\tilde{M}^{-1}_{poly} + \gamma_5\tilde{M}'^{-1}_{eig}$. However, this presents a possible conflict since $\tilde{M}^{-1}_{poly}$ will overlap on the deflated Hermitian eigenvector space. In order to prevent this, we also remove low eigenmode information from $\tilde{M}^{-1}_{poly}$. Since $\tilde{M}_{poly}$ is not Hermitian, the procedure is to define $\tilde{M}'_{poly} = \tilde{M}_{poly}\gamma_5$ and remove its overlapping Hermitian eigenvalue information using the eigenvectors from M'. Following the idea in Ref. [3], we define

$$e'^{(q)\dagger}_R\tilde{M}'^{-1}_{poly}e'^{(q)}_R \equiv \frac{1}{\xi'^{(q)}}, \quad (1.70)$$

where $e'^{(q)}_R$ is the eigenmode of M' generated within HFES method and $1/\xi'^{(q)}$ are the approximate eigenvalues of $\tilde{M}'^{-1}_{poly}$. The trace takes the following form:

$$Tr\left(\Theta M^{-1}\right) = \frac{1}{N}\sum_{n}^{N}\left(\eta^{(n)\dagger}\left[\Theta x^{(n)} - \Theta \tilde{x}'^{(n)}_{eig} - \left(\Theta \tilde{x}^{(n)}_{poly} - \Theta \tilde{x}'^{(n)}_{eigpoly}\right)\right]\right)$$

$$+ Tr\left(\Theta \gamma_5 \tilde{M}'^{-1}_{eig}\right) + Tr\left(\Theta \tilde{M}^{-1}_{poly} - \Theta \gamma_5 \tilde{M}'^{-1}_{eigpoly}\right), \qquad (1.71)$$

where

$$\tilde{x}'^{(n)}_{eigpoly} \equiv \gamma_5 \tilde{M}'^{-1}_{eigpoly}\eta^{(n)} = \gamma_5 \sum_{q}^{Q}\frac{1}{\xi'^{(q)}}e'^{(q)}_{R}\left(e'^{(q)\dagger}_{R}\eta^{(n)}\right). \qquad (1.72)$$

$\tilde{x}'^{(n)}_{eig}$ and $\tilde{x}^{(n)}_{poly}$ are defined in previous sections and

$$\tilde{M}'^{-1}_{eigpoly} \equiv \tilde{V}'_{R}\,\Xi^{-1}\,\tilde{V}'^{\dagger}_{R}, \qquad (1.73)$$

where $\tilde{V}'_{R}$ is defined above also and Ξ^{-1} is the diagonal matrix of size Q that contains approximate inverse eigenvalues, $1/\xi'^{(q)}$.

In the case of HFPS, $\tilde{M}^{-1}_{poly}$ is replaced by $\tilde{M}^{-1}_{pert}$ and all the calculations are repeated.

1.6   Operators Under Simulation

Expectation value of any operator $\mathcal{O}$ is given as

$$\langle \bar{\psi}\mathcal{O}\psi \rangle = -Tr(\mathcal{O}M^{-1}). \qquad (1.74)$$

Calculation of these values depends on the operator by varying degree because in some operators noise degrades the signal significantly, while in some cases noise does not impair the signal that much. Also, even though each operator is calculated with a real and imaginary part, only the real or imaginary term survives. This is because the quark propagator identity $S = \gamma_5 S^{\dagger}\gamma_5$, depending on the operator, allows only real or imaginary term to be nonzero on a given configuration for each space-time point.

In the simulation, though, quark propagator is not exact within noise techniques. This prevents the zero terms not to be exactly zero. From the theory, we know that these terms are purely noise and can just drop them from the calculations. This dropping does not bias the variance calculation but at the same time reduces variance.

In this dissertation, we present the results for nine different operators which can be divided into three different groups. Allowing μ to range from 1 to 4, three groups transform into nine different operators. Symbols and the equations related are shown in Table 1.2.

Table 1.2 Calculated operators and their representations

Name	Representation	Total operators
Scalar	$Re[\bar{\psi}(x)\psi(x)]$	1
Local vector	$Im[\bar{\psi}(x)\gamma_\mu\psi(x)]$	4
Point-split vector	$\kappa Im\left[\bar{\psi}(x+a_\mu)(1+\gamma_\mu)U_\mu^\dagger(x)\psi(x)\right]$	4
	$-\kappa Im\left[\bar{\psi}(x)(1-\gamma_\mu)U_\mu(x)\psi(x+a_\mu)\right]$	

1.7 Results

The goal of this work is to develop new methods and test if statistical error bar becomes smaller than the current noise subtraction techniques. We want to see if the methods developed by our group outperform tried and tested techniques, especially the perturbative subtraction (PS) method. Interestingly, we found some encouraging results which we will explain in this section.

In this dissertation, we first worked on Hermitian forced eigenspectrum noise subtraction (HFES) method for the QCD matrix. Then we combined HFES method with perturbative subtraction (PS) and polynomial subtraction (POLY) methods to form Hermitian forced perturbative subtraction (HFPS) and Hermitian forced polynomial subtraction (HFPOLY) methods, respectively. Also, we calculated the error associated with eigenspectrum subtraction (ES) and non-subtracted (NS) methods for comparison. We found three of the methods developed, namely POLY, HFPS, and HFPOLY, outperform PS method. Out of these methods, reduction of variance is found to be most effective for HFPOLY method. This result was consistent when the error was compared across different kappa values, lattice sizes, and operators.

We used two different programming languages, to complete this simulation. One is written in MATLAB and the other one in FORTRAN 90. MATLAB was used to simulate the smaller lattice of size 8^4 where we first saw our methods are useful. However, in MATLAB, we could not work on actual lattice sizes that are much bigger of size $24^3 \times 32$ and switched to FORTRAN 90, Randy Lewis' QQCD program.

Initially, we started our work with MATLAB and its "Parallel Computing Toolbox." This way we could run the code segments parallelly and obtain results much faster. The gauge configurations used in MATLAB were generated from the QQCD program. Due to the memory constraints 8^4 is the largest size of lattice that could be simulated in MATLAB. Also, algorithms written in MATLAB run slower than in FORTRAN 90 algorithms because even–odd preconditioning was not used on the former.

Because of the ease of testing various implementations and ability to present results quickly, we started with MATLAB algorithms to begin with debugging. After getting the knowledge that error bar is indeed decreasing we looked if we could see exactly the same result in QQCD program. After working for a while on the code in FORTRAN 90, we observe QQCD program and MATLAB program yielded the same result for 8^4 lattice. This made us more confident in our calculations. We then increased the size of lattice to $24^3 \times 32$ and analyzed the error reduction in more operators.

In MATLAB, we could only work with five different local operators as these operators respond best to unpartitioned noise. Unpartitioned noise spreads out across all degrees of freedom, whereas partitioned noise is only used across a specific degree of freedom, for example, color or time. Four local vector operators γ_μ and scalar operator $\bar{\psi}\psi$ work well with unpartitioned noise. However, nonlocal operators such as point-split operator are difficult to implement on MATLAB and are simulated only in QQCD program.

If the error bar gets smaller, then we would want to see how many subtracted eigenmodes are required to get the desired precision. Also, we want to do it for different kappa values and compare the error bars between them. Each simulation for a given gauge configuration and value of κ is conducted by averaging 200 $Z(4)$ noises. In order to gather an average error across gauge configurations, the errors are averaged over simulations.

Each simulation was conducted for three different values of κ, where $\kappa = \{0.1550, 0.1560, 0.1570\}$ for the small lattice of size 8^4. However, when we switched to larger matrix of size $24^3 \times 32$, our algorithms break down for $\kappa = 0.1570$. GMRES-DR algorithm failed to decrease the residual for linear equations as the number of iterations increased. Our group has now overcome this problem and has confirmed that deflation methods improve as the quark mass is increased.

Nevertheless, we were able to successfully reduce the error bars for the actual sized lattice QCD matrices for κ values 0.1550 and 0.1560. We noticed that as κ approaches kappa critical, eigenvalues start to get close to zero and it becomes tougher for the GMRES-DR algorithm to provide solution. The bigger the matrix, the tougher it gets. However, one interesting thing we observed is, HFES method provides a steeper decrease in error bar as kappa gets increased to kappa critical even though getting that smallest eigenvalue might be tougher. This fact is observed in the increased relative efficiency of the methods as κ increases towards κ_{crit} which we will observe from the figures to follow.

In the trace error-bar results that we are about to present, each graph represents operator/hopping parameter pair. The graph shows every method compared against each other plotted as the statistical error versus number of subtracted eigenmodes. By doing this, we wanted to see which method produces the least amount of error.

Each graph shows the "vanilla method" as the reference to compare. If a method cannot produce error smaller than this method, then that method is not effective at all. The legend on each graph is interpreted in Table 1.3.

Now we will analyze the progression of the error for different lattice sizes as the quark mass changes. Figures 1.5, 1.6, and 1.7 show the progression of local vector with $\mu = 4$, also called charge density, for 8^4 lattice, whereas Figs. 1.8 and 1.9 show the same progression for $24^3 \times 32$ lattice. In other words, Figs. 1.5, 1.6, and 1.7 represent the error calculations for κ values as 0.1550, 0.1560, and 0.1570, respectively, for a 8^4 lattice and Figs. 1.8 and 1.9 represent error calculations for κ

Table 1.3 The legend on each graphs and their representation

Symbol	Method	Representation
NS	No subtraction	Dark blue
ES	Eigenspectrum subtraction	Dotted green
HFES	Hermitian forced eigenspectrum subtraction	Magenta
PS	7th Order perturbative subtraction	Red
POLY	7th Order polynomial subtraction	Solid green
HFPS	HFES and PS combination	Black
HFPOLY	HFES and POLY combination	Pink

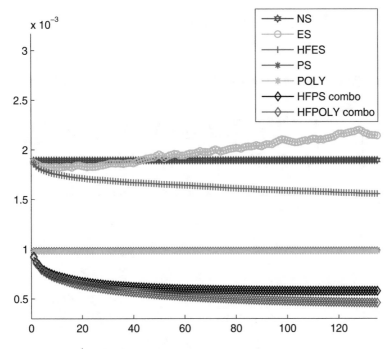

Fig. 1.5 Quenched 8^4 lattice simulation results for J_4 *charge density* at $\kappa = 0.1550$

values as 0.1550 and 0.1560, respectively, for $24^3 \times 32$ lattice for a local vector in fourth direction. We will present results only for the three operators local J_4, point split (or nonlocal) J_4, and scalar operators since the spatial J_i operators ($i = 1,2,3$) behaved much like the J_4 ones.

Eigenspectrum subtraction method produces error bar that is even greater than vanilla method. It shows no sign of decreasing variance as the error bar continues to climb up rather than going down as the number of deflated eigenvectors increase. As seen in the earlier works [4], ES method does not work because it fails to produce the eigenvectors that are perpendicular to each other. This method would work only if we used all the eigenvectors of the original matrix, however, that increases the

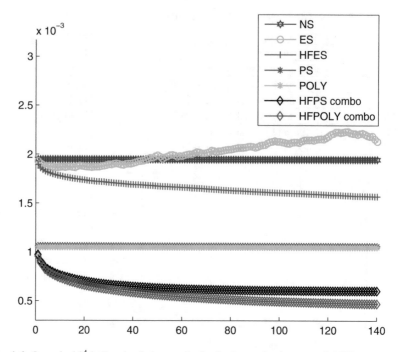

Fig. 1.6 Quenched 8^4 lattice simulation results for J_4 *charge density* at $\kappa = 0.1560$

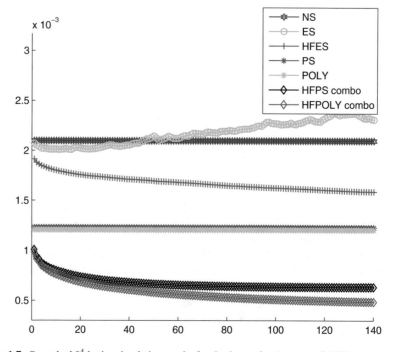

Fig. 1.7 Quenched 8^4 lattice simulation results for J_4 *charge density* at $\kappa = 0.1570$

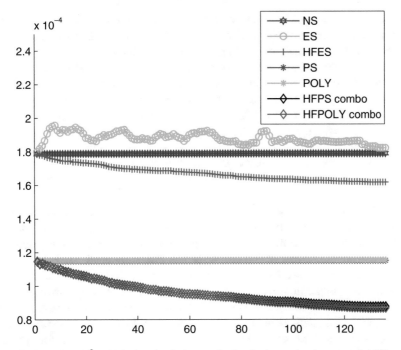

Fig. 1.8 Quenched $24^3 \times 32$ lattice simulation results for J_4 *charge density* at $\kappa = 0.1550$

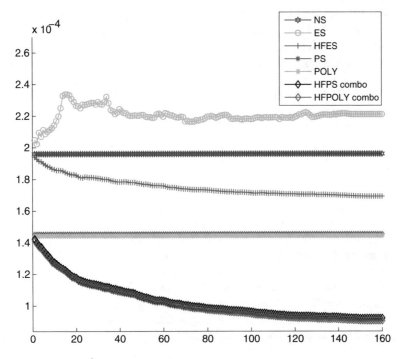

Fig. 1.9 Quenched $24^3 \times 32$ lattice simulation results for J_4 *charge density* at $\kappa = 0.1560$

matrix vector multiplications to a point that it becomes unuseful. It will be apparent after looking all the figures that ES method does not work for all lattice sizes, mass values, and operators and hence will not be discussed further.

Hermitian forced eigenspectrum subtraction (HFES) method which is built modifying the ES method reduces the error bar significantly. As the number of deflated eigenvectors is increased, variance associated with the simulation decreases. If we compare HFES results from Figs. 1.5, 1.6, and 1.7, plot of HFES error begins from the NS method for $\kappa = 0.1550$, while for κ value 0.1560 and 0.1570 the beginning point of HFES keeps shifting away from non-subtracted method. This was observed when the smallest eigenvalue was significantly smaller than second smallest eigenvalue. Since smallest eigenvalues of QCD matrix contribute greatest to the trace, this shifting occurs.

We observed that, on average, one out of ten gauge configurations of size 8^4 has smallest and second smallest eigenvalues that are different drastically especially so for κ_{cric}. So, for the 8^4 lattice, we generated one hundred configurations and averaged over the errors. This shifting is not that significant for larger lattices however. So, we assumed ten configurations for large matrices should be enough.

We can see similar improvement in error calculations for larger matrices in HFES method also. Actually, we can see that POLY method is almost comparable to the PS method, while the combination methods HFPS and HFPOLY outdo the perturbation method. As we pointed out previously, we have not shown the figures for operators in one to three directions because they are almost equal to the errors for the operators in fourth direction.

Let us talk about the progression of error in case of nonlocal operators in fourth direction, also called point-split operators. Similar to the local vectors, first three nonlocal vectors also have similar trace properties to the fourth nonlocal vector. So, we only show the figures for the fourth one. Figures 1.10, 1.11, and 1.12 show the corresponding error bar for various methods calculated with kappa values 0.1550, 0.1560, and 0.1570, respectively, for a smaller lattice. HFPOLY is the best method for point-split operator for larger lattices as well (Figs. 1.13 and 1.14).

For the scalar operator also HFPOLY seems to be the best method. For the smaller lattice though, it is clearly seen that deflation of about 15 eigenmodes is more than sufficient. After that actually it plateaus out if not increases slightly. Again, like in the other operators, because of the same reason as explained earlier, beginning point of HFES, HFPS, and HFPOLY is not from NS, PS, and POLY method, respectively. One interesting thing to note here is that for $\kappa = 0.1570$ and 8^4 lattice, HFES method is performing better than PS method even after deflation of 15 eigenmodes. Similarly, for $\kappa = 0.1560$ and larger lattice HFES method is outperforming PS method. Actually this time, it is outperforming POLY method also. For smaller kappa values $\kappa = 0.155$ though, HFES cannot outperform PS method. We know from the theory that PS method suits better for smaller kappa values, whereas HFES method is better for smaller eigenvalues, i.e., when kappa is close to critical value which is usually larger. So, it makes sense that as kappa is increasing HFES is getting better than PS method.

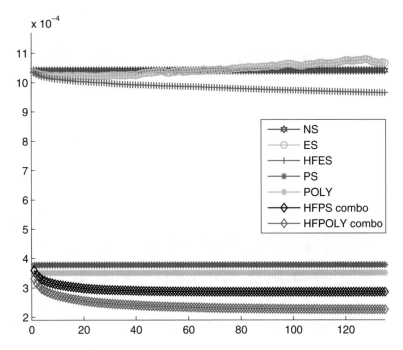

Fig. 1.10 Quenched 8^4 lattice simulation results for *point-split vector*$_4$ at $\kappa = 0.1550$

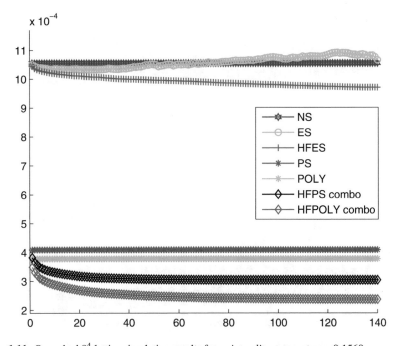

Fig. 1.11 Quenched 8^4 lattice simulation results for *point-split vector*$_4$ at $\kappa = 0.1560$

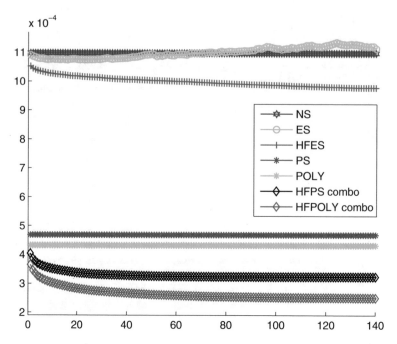

Fig. 1.12 Quenched 8^4 lattice simulation results for *point-split vector*$_4$ at $\kappa = 0.1570$

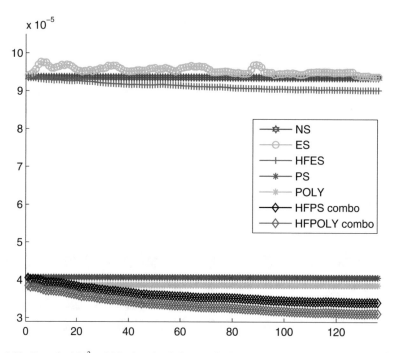

Fig. 1.13 Quenched $24^3 \times 32$ lattice simulation results for *point-split vector*$_4$ at $\kappa = 0.1550$

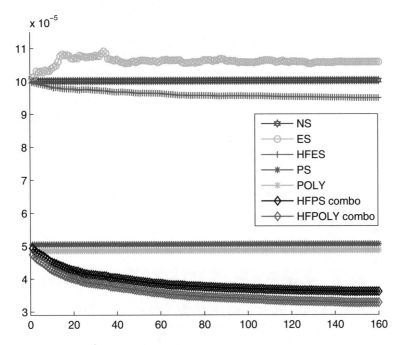

Fig. 1.14 Quenched $24^3 \times 32$ lattice simulation results for *point-split vector$_4$* at $\kappa = 0.1560$

We have calculated the relative efficiency between the PS and HFPOLY method to get an idea of which method is better at what mass, operator, and kappa values. We define the relative efficiency, RE, of the two methods as

$$RE \equiv \left(\frac{1}{\delta y^2} - 1 \right) \times 100, \qquad (1.75)$$

where δy is the relative error bar (Figs. 1.15, 1.16, 1.17, 1.18 and 1.19).

Tables 1.4, 1.5, 1.6, 1.7, and 1.8 show the relative efficiency as calculated for different parameters. Here, we compare our methods with PS method. We would see, three of the methods POLY, HFPS, and HFPOLY outperform PS method across all lattice size and operators. To compare the efficiency with non-subtracted techniques, we have added a column for NS also. Comparison of relative efficiency across operators shows HFES, HFPS, and HFPOLY all get better as kappa increases. Negative sign when PS and HFES are compared indicates HFES is still not better than PS in most of the cases. However, in Table 1.6, we can see HFES performs better than PS method. At $\kappa = 0.1570 = \kappa_{cric}$, when eigenvalues are close to zero, inverse of a matrix obtained gets more accurate and hence HFES becomes better. One can expect similar trend for larger matrices also.

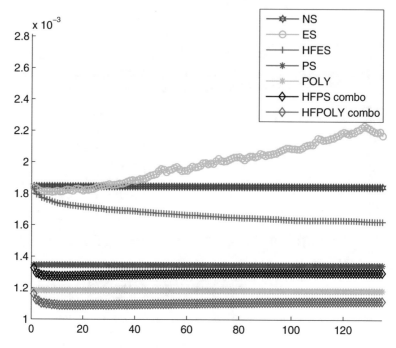

Fig. 1.15 Quenched 8^4 lattice simulation results for *scalar* at $\kappa = 0.1550$

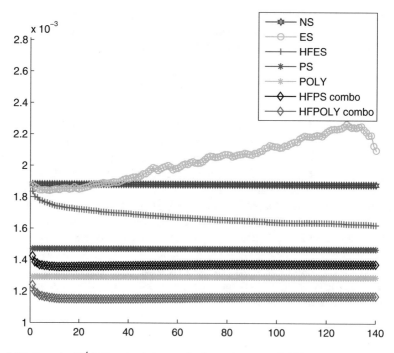

Fig. 1.16 Quenched 8^4 lattice simulation results for *scalar* at $\kappa = 0.1560$

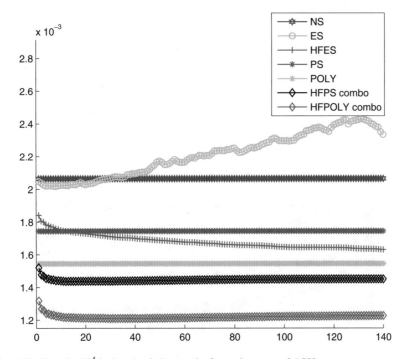

Fig. 1.17 Quenched 8^4 lattice simulation results for *scalar* at $\kappa = 0.1570$

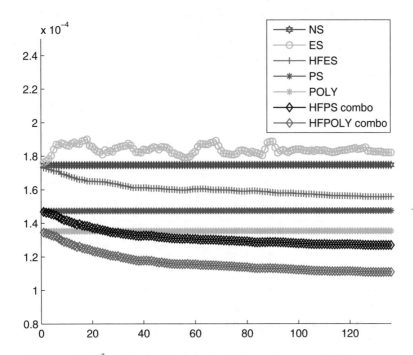

Fig. 1.18 Quenched $24^3 \times 32$ lattice simulation results for *scalar* at $\kappa = 0.1550$

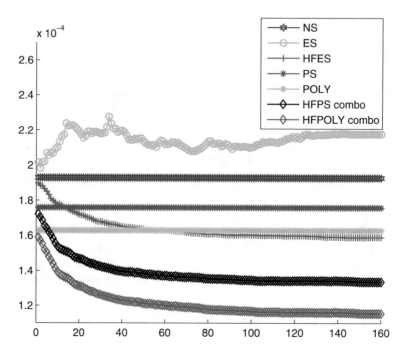

Fig. 1.19 Quenched $24^3 \times 32$ lattice simulation results for *scalar* at $\kappa = 0.1560$

Table 1.4 Comparison of relative efficiency for operators on the 8^4 lattice at $\kappa = 0.1550$

	Scalar		Local J_4		Nonlocal J_4	
Subtraction	*vs.* NS	*vs.* PS	*vs.* NS	*vs.* PS	*vs.* NS	*vs.* PS
POLY	143.4%	29.4%	273.6%	1.8%	778.3%	15.7%
HFES	29.9%	−30.9%	48.7%	−59.5%	16.6%	−84.7%
HFPS	101.8%	7.3%	977.5%	193.5%	1226.7%	74.7%
HFPOLY	173.2%	45.3%	1623.0%	369.3%	2033.6%	180.9%

Table 1.5 Comparison of relative efficiency for operators on the 8^4 lattice at $\kappa = 0.1560$

	Scalar		Local J_4		Nonlocal J_4	
Subtraction	*vs.* NS	*vs.* PS	*vs.* NS	*vs.* PS	*vs.* NS	*vs.* PS
POLY	112.3%	29.8%	244.8%	2.8%	678.4%	17.1%
HFES	34.1%	−18.0%	53.7%	−54.2%	18.0%	−82.2%
HFPS	87.1%	14.4%	937.1%	209.2%	1102.3%	80.8%
HFPOLY	157.5%	57.5%	1608.3%	409.3%	1875.5%	197.1%

For the larger lattice at $\kappa = 0.155$, relative efficiency is not very different between operators as can be seen in Tables 1.7 and 1.8. As the kappa increases, relative efficiency is also getting better.

Table 1.6 Comparison of relative efficiency for operators on the 8^4 lattice at $\kappa = 0.1570$

Subtraction	Scalar		Local J_4		Nonlocal J_4	
	vs. NS	*vs.* PS	*vs.* NS	*vs.* PS	*vs.* NS	*vs.* PS
POLY	78.6%	27.4%	201.6%	3.6%	546.7%	17.6%
HFES	60.5%	14.5%	74.2%	−40.13%	26.0%	−77.1%
HFPS	102.5%	44.5%	992.7%	275.5%	1051.5%	109.3%
HFPOLY	183.5%	102.2%	1755.5%	537.5%	1834.6%	251.7%

Table 1.7 Relative efficiency for operators on the $24^3 \times 32$ lattice at $\kappa = 0.1550$

Subtraction	Scalar		Local J_4		Nonlocal J_4	
	vs. NS	*vs.* PS	*vs.* NS	*vs.* PS	*vs.* NS	*vs.* PS
POLY	66.5%	18.5%	140.8%	−0.53%	485.6%	10.7%
HFES	25.7%	−10.5%	22.4%	−49.4%	7.5%	−79.6%
HFPS	89.5%	34.9%	316.4%	72.0%	652.6%	42.3%
HFPOLY	148.7%	77.6%	325.7%	75.9%	799.6%	70.1%

Table 1.8 Relative efficiency for operators on the $24^3 \times 32$ lattice at $\kappa = 0.1560$

Subtraction	Scalar		Local J_4		Nonlocal J_4	
	vs. NS	*vs.* PS	*vs.* NS	*vs.* PS	*vs.* NS	*vs.* PS
POLY	40.1%	16.6%	84.4%	1.3%	324.5%	7.7%
HFES	47.2%	22.5%	34.2%	−26.2%	11.6%	−71.6%
HFPS	107.9%	73.0%	352.09%	148.4%	671.3%	95.7%
HFPOLY	178.7%	131.9%	367.8%	157.05%	839.5%	138.4%

We expect efficiency to improve further as we move on towards lower quark masses. However, as we increased kappa from 0.1560 to 0.1570, GMRES-DR failed to converge. This is because of the eigenvalues that are very close to zero. This region where poles occur is challenging and most of the lattice QCD algorithms fail at this point. However, one positive that we can take from our results is that if we can get the smallest eigenvalues, we can use them for deflation to converge GMRES-DR and also use them to construct the trace of the matrix. We expect relative efficiency gained with this method would be huge. Our latest results indeed confirm this expectation. Note that deflation applied to hierarchical probing has been obtained in Ref. [11].

1.8 Conclusions

Our polynomial and perturbative deflation combination methods produce very encouraging results. We expect efficiency to improve further as we move on towards lower quark masses. As we increased kappa from 0.1560 to 0.1570, efficiency was significantly increased across all the lattice sizes where we could complete our calculation. At critical kappa value for large matrices, we had a difficult time to get GMRES-DR to converge because of the eigenvalues very close to zero. Our group has now overcome this issue and used MinRes algorithm to compute such small eigenvalues. We can then use them for deflation to converge GMRES-DR and also use them to construct the trace of the matrix. We expect relative efficiency gained with this method would be huge. We are hopeful that our methods will be important in QCD calculations.

Acknowledgements We thank the University Research Committee of Baylor University for their support of this project. We thank Abdou Abdel-Rehim and Victor Guerrero for their contribution to noise subtraction programming, and Randy Lewis for the use of his QQCD program. Also, we would like to thank Carlton DeTar and Doug Toussaint who made MILC collaboration lattices available. We would like to thank the Texas Advanced Computing Center for account support. Finally, we would like to thank Everest Institute of Science and Technology (EVIST), Kathmandu, Nepal and Neural Innovations LLC, Hewitt, Texas for assistance during research.

References

1. W. Wilcox, *Numerical Challenges in Lattice Quantum Chromodynamics*. Lecture Notes in Computer Science and Engineering, vol. 15 (Springer, Berlin, 2000), p. 127
2. S. Bernardson, P. McCarty, C. Thron, Comput. Phys. Commun. **78**, 256 (1993)
3. V. Guerrero, R.B. Morgan, W. Wilcox, PoS (LAT2009) 041
4. V. Guerrero, Doctoral Dissertation, 2011, http://hdl.handle.net/2104/8214
5. S. Baral, W. Wilcox, R. Morgan, arXiv:1611.01536[Hep-lat]
6. S. Baral, https://hdl.handle.net/2104/10503
7. Q. Liu, W. Wilcox, R. Morgan, arXiv:1405.1763 [hep-lat]
8. R.B. Morgan, SIAM J. Sci. Comput. **24**, 20 (2002)
9. D. Darnell, R.B. Morgan, W. Wilcox, Linear Algebra Appl. **429**, 2415 (2008)
10. W. Wilcox, PoS (LAT 2007) 025
11. A. Gambhir, A. Stathopoulos, K. Orginos, arXiv:1603.05988 [hep-lat]

Chapter 2
TF Model for Mesonic Matters

2.1 Introduction

Lattice QCD is a very important tool used by particle physicists to investigate the properties of baryons and mesons. Lattice techniques [1–5] are presently being employed to understand and elucidate the pentaquark and tetraquark observations made by Belle [6–9], BESIII [10, 11], and LHCb [12–16]. However, as the quark content increases, it becomes computationally expensive and time-intensive to do the lattice calculations. Every state must be investigated separately, which means a great deal of analysis on Wick contractions and specialized computer coding. In addition, as one adds more quarks, the states will become larger and the lattice used must also increase in volume. There is therefore a need for reliable quark models that can give an overview of many states to help guide these expensive lattice calculations. The MIT bag model [17–19] and Nambu-Jona-Lasinio [20] model are two of these quark models. Another approach, the Thomas-Fermi (TF) statistical model has been amazingly successful in the explanation of atomic spectra and structure, as well as nuclear applications. Our group has adopted the TF model and applied it to collections of many quarks [21–23]. One advantage our model has over bag models is the inclusion of nonperturbative Coulombic interactions. One would expect that the TF quark model would become increasingly accurate as the number of constituents is increased, as a statistical treatment is more justified. The main usefulness will be to see systematic trends as the parameters of the model are varied. It could also be key to identifying *families* of bound states, rather than individual cases. The TF quark model has already been used to investigate multi-quark states of baryons [24]. In this paper we have extended the TF quark model to mesonic states in order to investigate the stability of families built from some existing mesons and observed new exotic states, concentrating on heavy-light quark combinations. Although our model is nonrelativistic, we will see that this assumption is actually numerically consistent as quark content is increased. We have not yet included spin–spin interactions in our meson model, and is not as developed as our previous baryon

© Springer Nature Switzerland AG 2019
S. Baral, *Thomas-Fermi Model for Mesons and Noise Subtraction Techniques in Lattice QCD*, Springer Theses, https://doi.org/10.1007/978-3-030-30904-6_2

model. Because of this we cannot yet make predictions for energies of multi-quark states, although we can make observations regarding family stability.

Our paper is organized as follows: In Sect. 2.2 we will define the energies of the model in terms of the TF density of states. In Sect. 2.3 we will examine the classical color couplings in mean field theory and systematically determine the probabilities of interactions for quark–quark, quark–antiquark, and antiquark–antiquark interactions. We obtain the system energies in Sect. 2.4, the TF quark equations in Sect. 2.5, and recharacterize the model in terms of new dimensionless variables in Sect. 2.6. The application of the model to heavy quark–light quark mesonic matter is initiated in Sect. 2.7, where we define three types of multi-quark families involving charm quarks: charmonium ("Case 1"), Z-meson type ("Case 2"), and D-meson type ("Case 3"). These states as well as their bottom-quark analogs are constructed in Sect. 2.7, where we examine the energy slopes to determine family stability. Numerical results and discussions are presented in Sects. 2.8 and 2.9.

2.2   The TF Meson Model

The TF statistical model is a semiclassical quantum mechanical theory developed for many-fermion systems. In this model the assumption is made that quarks are distributed uniformly in each volume element ΔV, while at the same time the quark density $n_q(r)$ can vary from one small volume element to the next. For a small volume element ΔV, and for the system of quarks in its ground state, we can fill out a spherical momentum space volume V_F up to the Fermi momentum p_F, and thus

$$V_F \equiv \frac{4}{3}\pi p_F^3(\vec{r}).\tag{2.1}$$

The corresponding phase space volume is

$$\Delta V_{ph} \equiv V_F \Delta V = \frac{4}{3}\pi p_F^3(\vec{r})\Delta V.\tag{2.2}$$

Let us say g^I is the degeneracy of a quark flavor I. Then quarks in ΔV_{ph} are distributed uniformly with g^I quarks per h^3 of this phase space volume, where h is Planck's constant. The number of quarks in ΔV_{ph} is

$$\Delta N_{ph} \equiv \frac{g^I}{h^3}\Delta V_{ph} = \frac{4\pi g^I}{3h^3}p_F^3(\vec{r})\Delta V.\tag{2.3}$$

The number of quarks in ΔV is

$$\Delta N \equiv n_q(\vec{r})\Delta V,\tag{2.4}$$

where $n_q(\vec{r})$ is the quark density. Equating Eqs. (2.3) and (2.4), we obtain

$$n_q(\vec{r}) = \frac{4\pi g^I}{3h^3} p_F^3(\vec{r}). \tag{2.5}$$

The fraction of quarks at position $\vec{r}$ that have momentum between p and $p + dp$ in spherical momentum space is

$$F_{(\vec{r})}(p)dp = \begin{cases} \frac{4\pi p^2 dp}{\frac{4}{3}\pi p_F^3(\vec{r})}, & \text{if } p \le p_F(\vec{r}), \\ 0, & \text{otherwise.} \end{cases} \tag{2.6}$$

Using the classical expression for kinetic energy of a quark with mass m_I, the kinetic energy per unit volume for a system of quarks is

$$t(\vec{r}) = \int^{p_F} \frac{p^2}{2m_I} n_q(\vec{r}) F_{(\vec{r})}(p)dp = C_F \left(n_q(\vec{r})\right)^{5/3}, \tag{2.7}$$

where

$$C_F = \frac{\left(6\pi^2 \hbar^3\right)^{5/3}}{20\pi^2 \hbar^3 m_I \left(g^I\right)^{2/3}}. \tag{2.8}$$

Equation (2.7) shows that kinetic energy per volume is proportional to the 5/3-rd power of quark density. To obtain the total kinetic energy (T), we will have to integrate this expression over all the spatial volume:

$$T = \int t(\vec{r}) d^3r. \tag{2.9}$$

On the other hand, the total potential energy (U) due to the interactions of system of quarks with one another is given by

$$U = V_N \int \int \frac{n_q(\vec{r})n_q(\vec{r}\,')}{|\vec{r} - \vec{r}\,'|} d^3r \, d^3r', \tag{2.10}$$

where V_N is a factor depending on the type of interaction between quarks. It will be explained in the sections to follow. The total energy for a system of quarks is therefore

$$E = T + U = C_F \int \left(n_q(\vec{r})\right)^{5/3} d^3r + V_N \int \int \frac{n_q(\vec{r})n_q(\vec{r}\,')}{|\vec{r} - \vec{r}\,'|} d^3r d^3r'. \tag{2.11}$$

In order to minimize energy while keeping the number of quarks constant, we can add a Lagrange multiplier term of the form $\lambda \left(\int n_q(\vec{r}) d^3(r) - N_q \right)$ to the expression for total energy, where N_q is the fixed total number of quarks in the given system.

A given quark actually has a definite color and flavor. In the sections to follow we will replace the number density of quarks, $n_q(\vec{r})$, by the quantity $n_i^I(\vec{r})$, where color is represented by index i and flavor by index I. It will then be summed over flavor and color indices to account for the energies of all of the quarks in the given system.

2.3   Residual Coulombic Coupling and Interaction Probabilities

The color couplings of quarks and antiquarks in our model originate from the Coulombic interactions expected at the classical level [25]. In the following we define η to be the number of quark/antiquark pairs in the meson, which is assumed to be a color singlet. In addition, g represents the strong coupling constant.

The types of interactions between the particles can then be categorized as:

Color–Color Repulsion (CCR): Interactions between quarks with same colors are repulsive with coupling constant $4/3g^2$. The interactions are red-red (rr), green-green (gg), and blue-blue (bb).

Color–Color Attraction (CCA): Interactions between quarks with different colors are attractive with coupling constant $-2/3g^2$. The interactions are rb, rg, bg, br, gr, and gb.

Color–Anticolor Repulsion (CAR): Interactions between quarks and antiquarks with different color/anticolors are repulsive with coupling constant $2/3g^2$. The interactions are $r\bar{b}$, $r\bar{g}$, $b\bar{g}$, $b\bar{r}$, $g\bar{r}$, and $g\bar{b}$.

Color–Anticolor Attraction (CAA): Interactions between quarks and antiquarks with same color/anticolors are attractive with coupling constant $-4/3g^2$. The interactions are $r\bar{r}$, $b\bar{b}$, and $g\bar{g}$.

Anticolor–Anticolor Repulsion (AAR): Interactions between antiquarks with same anticolors are repulsive with coupling constant $4/3g^2$. The interactions are $\bar{r}\bar{r}$, $\bar{b}\bar{b}$, and $\bar{g}\bar{g}$.

Anticolor–Anticolor Attraction (AAA): Interactions between antiquarks with different anticolors are attractive with coupling constant $-2/3g^2$. The interactions are $\bar{r}\bar{b}$, $\bar{r}\bar{g}$, $\bar{b}\bar{g}$, $\bar{b}\bar{r}$, $\bar{g}\bar{r}$, and $\bar{g}\bar{b}$.

When η pairs of quarks interact with each other, the six types of color interactions appear with different probabilities. So, we need to find a way to count and formulate color-interaction probabilities in order to apply the TF model for mesonic matter.

Each of the η pairs must be a color singlet and each singlet can be achieved in three different ways, red-antired, green-antigreen, and blue-antiblue. We will term each quark–antiquark pair a "pocket." Out of η pockets let us say we have x number of red-antired, y number of blue-antiblue, and $z = \eta - x - y$ number of green-antigreen combinations which will be represented by R, B, and G, respectively. This is depicted in Table 2.1.

Table 2.1 Color singlet counting and representation

Number of pockets	Type of pocket	Representation
x	$r\bar{r}$	R
y	$b\bar{b}$	B
z	$g\bar{g}$	G

Table 2.2 Color interactions for same pockets, different pockets, and within a pocket

RR	r	$\bar{r}$	RB	r	$\bar{r}$	R	r
r	rr	$r\bar{r}$	b	br	$b\bar{r}$	$\bar{r}$	$\bar{r}r$
$\bar{r}$	$\bar{r}r$	$\bar{r}\bar{r}$	$\bar{b}$	$\bar{b}r$	$\bar{b}\bar{r}$		

Table 2.3 Color interaction counting and indices

Index(i)	Interaction type	$(RR)_i$	$(RB)_i$	$(R)_i$
1	CCR	1	0	0
2	CCA	0	1	0
3	CAR	0	2	0
4	CAA	2	0	1
5	AAR	1	0	0
6	AAA	0	1	0

When two R pockets interact, for example, there are four color combinations as depicted in the left box of Table 2.2. Out of these, $r\bar{r}$ and $\bar{r}r$ belong to interaction type CAA, rr to CCR, and $\bar{r}\bar{r}$ to AAR. The middle box RB depicts color combinations when different types of pocket interact. The rightmost box R represents a color combination within this pocket which means there is interaction type CAA only. These color combinations are counted and depicted in Table 2.3. Of course BB and GG have the same interaction numbers as in the left box. In addition, BG and GR have same interaction numbers as in the middle box.

In Table 2.3 the six interaction types are given an index. The last three columns give the number of interactions contained in the previous table.

There are $x(x-1)/2$ number of ways the RR type of interaction takes place. Similarly, $y(y-1)/2$ and $z(z-1)/2$ are the number of times BB and GG happen. Also, RB, BG, and RG type of interactions occur xy, yz, and xz times, respectively.

In the next step we varied x from 0 to η and y from 0 to $\eta-x$, thereby giving equal footing to all the color combinations and counted all the possible color interactions, E_i. The corresponding expression is given in the equation below.

$$E_i = \sum_{x=0}^{\eta} \sum_{y=0}^{\eta-x} \frac{\eta!}{x!y!z!} \tag{2.12}$$

$$\times \left\{ \frac{x(x-1) + y(y-1) + z(z-1)}{2} \right.$$

$$\left. \times (RR)_i + (xy + yz + zx)\,(RB)_i + \eta(R)_i \right\}.$$

Here E_i refers to total occurrence of the ith color coupling. For example, E_1 refers to total number of times the AAA type of interaction occurs out of $3^\eta \times \eta(2\eta - 1)$ possibilities.

To understand this equation, let us break it into two parts. First, if we take the terms inside the curly braces and add over the indices, we get $\eta(2\eta - 1)$, i.e.,

$$\sum_i \left\{ \frac{x(x-1) + y(y-1) + z(z-1)}{2}(RR)_i + (xy + yz + zx)(RB)_i + \eta(R)_i \right\}$$

$$= \eta(2\eta - 1),$$

(2.13)

regardless of x, y, and z values as long as the constraint $x + y + z = \eta$ is satisfied. This is just the number of ways 2η quarks can be paired. Second, if we replace the curly braces by unity, we get 3^η, the total number of pocket combinations in a meson with η pairs of quarks. Thus, E_i gives the total occurrence of the ith color interaction out of $3^\eta \eta(2\eta - 1)$ possibilities.

After solving Eq. (2.12) using *Mathematica*, we obtained the outcomes, E_i. They are listed in Table 2.4. In the same table, we divided each event by $3^\eta \eta(2\eta - 1)$ and obtained probabilities for each type of interaction. In addition, the coupling strength, C_i, is defined for each interaction type.

If we add the product of coupling and probabilities from Table 2.4, we get $-\frac{4}{3}g^2/(2\eta - 1)$, very similar to the baryon case [24]. The negative sign indicates that the system is attractive because of the collective residual color coupling alone, even in the absence of volume pressure. This gives rise to a type of matter that is bound, but does not correspond to confined mesonic matter, as discussed in Ref. [21]. We are interested in confined matter and will need to add a pressure term to the energy to enforce this.

As a check on our counting, consider the total color charge of the quarks:

$$\vec{Q} = \sum_{i=1}^{2\eta} \vec{q}_i.$$

(2.14)

Table 2.4 The coupling strength, total outcome, and probabilities for all six interactions

Symbol (C)	Couplings	Symbol (E)	Number of outcomes	Symbol(P)	Probabilities
C_1	$\frac{4}{3}g^2$	E_1	$3^{\eta-1}\frac{\eta(\eta-1)}{2}$	P_1	$\frac{\eta-1}{6(2\eta-1)}$
C_2	$-\frac{2}{3}g^2$	E_2	$3^{\eta-1}\eta(\eta-1)$	P_2	$\frac{\eta-1}{3(2\eta-1)}$
C_3	$\frac{2}{3}g^2$	E_3	$3^{\eta-1}2\eta(\eta-1)$	P_3	$\frac{2(\eta-1)}{3(2\eta-1)}$
C_4	$-\frac{4}{3}g^2$	E_4	$3^{\eta-1}\eta(\eta+2)$	P_4	$\frac{\eta+2}{3(2\eta-1)}$
C_5	$-\frac{4}{3}g^2$	E_5	$3^{\eta-1}\frac{\eta(\eta-1)}{2}$	P_5	$\frac{\eta-1}{6(2\eta-1)}$
C_6	$-\frac{2}{3}g^2$	E_6	$3^{\eta-1}\eta(\eta-1)$	P_6	$\frac{\eta-1}{3(2\eta-1)}$

Squaring on both sides, we get

$$
\begin{aligned}
\vec{Q}^2 &= \sum_{i=1}^{2\eta} \vec{q}_i^{\,2} + 2\sum_{i \neq j} \vec{q}_i \cdot \vec{q}_j \\
&= 2\eta \times \frac{4}{3}g^2 + 2 \times \frac{1}{3^\eta}\sum_i E_i C_i \\
&= 0.
\end{aligned}
\tag{2.15}
$$

This shows our model is an overall color singlet. Note that we divided by 3^η in the second term to average over all the possible configurations.

The TF quark model replaces the sum over particle number in particle interaction models with an integral over the density of state particle properties. We will weight interaction strengths, taken from the classical theory, by the probabilities of various interactions in the color sector, which we assume to be flavor independent. So, we need a connection between particle number and probability. We will assume as in other TF quark models that these interaction probabilities are proportional to the number of particle interaction terms. Also, treating color combinations on an equal footing, each color now shares one-third of the total probability. So, we divide the probabilities in Table 2.4 by three as shown in Table 2.5. Also, notice that we have generated a new way of representing probabilities using double color indices i and j and barred and double barred symbols for P. No bar is for color–color probabilities, one bar is for color–anticolor probabilities, and two bars are for anticolor–anticolor probabilities. From Table 2.5, we can see that we correctly obtain $\sum_{i \leq j}^3 P_{ij} + \bar{P}_{ij} + \bar{\bar{P}}_{ij} = 1$ for the sum of all the probabilities.

Table 2.5 Coupling constants and probabilities for quark–antiquark interactions in mesons

New symbol	Old symbol	Probability values
P_{ii}	$\frac{P_1}{3}$	$\dfrac{(\eta - 1)}{18(2\eta - 1)}$
$P_{ij}, i < j$	$\frac{P_2}{3}$	$\dfrac{\eta - 1}{9(2\eta - 1)}$
$\bar{P}_{ij}, i < j$	$\frac{P_3}{3}$	$\dfrac{2(\eta - 1)}{9(2\eta - 1)}$
$\bar{P}_{ii}$	$\frac{P_4}{3}$	$\dfrac{(\eta + 2)}{9(2\eta - 1)}$
$\bar{\bar{P}}_{ii}$	$\frac{P_5}{3}$	$\dfrac{(\eta - 1)}{18(2\eta - 1)}$
$\bar{\bar{P}}_{ij}, i < j$	$\frac{P_6}{3}$	$\dfrac{(\eta - 1)}{9(2\eta - 1)}$

2.4 System Energies and Equations

Our multi-quark system consists of an equal number of η quarks and η antiquarks. Each quark or antiquark carries both color and flavor. So, let us introduce N^I as the number of quarks with flavor index I, and $\bar{N}^I$ as number of antiquarks with anti-flavor index I such that

$$\sum_I g^I N^I = \eta, \tag{2.16}$$

and

$$\sum_I \bar{g}^I \bar{N}^I = \eta. \tag{2.17}$$

In terms of quark density these equations can be expressed as

$$\sum_I \int d^3r\, n_i^I(r) = \eta/3, \tag{2.18}$$

and

$$\sum_I \int d^3r\, \bar{n}_i^I(r) = \eta/3. \tag{2.19}$$

In Eqs. (2.18) and (2.19) degeneracy factors are already included in quark densities $n_i^I(r)$ and $\bar{n}_i^I(r)$.

Similarly, the sum over the color index gives the total number for a given flavor. Thus

$$\sum_i \int d^3r\, n_i^I(r) = N^I g^I \tag{2.20}$$

and

$$\sum_i \int d^3r\, \bar{n}_i^I(r) = \bar{N}^I \bar{g}^I. \tag{2.21}$$

Also, for the convenience, we will introduce the single-particle normalized density

$$\hat{n}_i^I \equiv \frac{3n_i^I}{N^I} \tag{2.22}$$

and

$$\hat{\bar{n}}_i^I \equiv \frac{3\bar{n}_i^I}{\bar{N}^I}. \tag{2.23}$$

This form of quark density will be helpful in correctly normalizing the TF interaction energy when continuum sources are used.

In the earlier section, we calculated probability for six types of interactions between colors. We will now see how these color interactions are associated with flavor numbers.

For a quark–antiquark system with 2η total number of particles, the number of interactions possible is $\eta(2\eta - 1)$. Out of these **CCA** and **CCR** types occur

$$\left(\sum_I \frac{N^I \left(N^I - 1 \right)}{2} (g^I)^2 + \sum_{I \neq J} \frac{N^I N^J}{2} g^I g^J + \sum_I N^I \frac{g^I(g^I - 1)}{2} \right)$$

times, **AAR** and **AAA** occur

$$\left(\sum_I \frac{\bar{N}^I \left(\bar{N}^I - 1 \right)}{2} (\bar{g}^I)^2 + \sum_{I \neq J} \frac{\bar{N}^I \bar{N}^J}{2} \bar{g}^I \bar{g}^J + \sum_I \bar{N}^I \frac{\bar{g}^I(\bar{g}^I - 1)}{2} \right)$$

times, and **CAR** and **CAA** occur

$$\left(\sum_{I,J} \bar{N}^I N^J \bar{g}^I g^J \right)$$

times. For a consistency check, we are going to add all of these seven terms. We begin with the three **CCA** and **CCR** terms, for which we have

$$\sum_I \frac{N^I \left(N^I - 1 \right)}{2} (g^I)^2 + \sum_{I \neq J} \frac{N^I N^J}{2} g^I g^J + \sum_I N^I \frac{g^I(g^I - 1)}{2} \tag{2.24}$$

$$= \sum_I \frac{N^I g^I}{2} \left(N^I g^I - 1 \right) + \sum_{I \neq J} \frac{N^I N^J}{2} g^I g^J$$

$$= \sum_I \frac{\left(N^I g^I \right)^2}{2} - \sum_I \frac{\left(N^I g^I \right)}{2} + \sum_{I \neq J} \frac{N^I N^J}{2} g^I g^J$$

$$= \frac{1}{2} \left(\sum_I N^I g^I \right)^2 - \sum_I \frac{\left(N^I g^I \right)}{2}$$

$$= \frac{1}{2} \left(\eta^2 - \eta \right).$$

The interaction between anti-flavors, i.e., the **AAR** and **AAA** terms, also yields $\eta^2/2 - \eta/2$. The last term involving one flavor and one anti-flavor gives η^2. Adding all of them yields $\eta(2\eta - 1)$ as it should.

We will now use this information to write expressions for kinetic and potential energies. Building up the expression from Sect. 2.2, we can write

$$T = \sum_{i,I} \int^{r_{\max}} d^3r \, \frac{\left(6\pi^2\hbar^3 N^I \hat{n}_i^I(r)\right)^{5/3}}{20\pi^2\hbar^3 m^I (g^I)^{2/3}} + \sum_{i,I} \int^{r_{\max}} d^3r \, \frac{\left(6\pi^2\hbar^3 \bar{N}^I \hat{\bar{n}}_i^I(r)\right)^{5/3}}{20\pi^2\hbar^3 \bar{m}^I (\bar{g}^I)^{2/3}},$$

(2.25)

for the kinetic energy from quarks and antiquarks. Here we have assumed that radius of the objects is finite. Similarly, assigning the probabilities and couplings from the interaction terms shown in Table 2.5, we can now define total potential energy to be

$$U = \frac{4}{3}g^2 \sum_I \frac{N^I (N^I - 1)}{2} \int\int d^3r \, d^3r'$$

$$\times \frac{\left(\sum_i P_{ii} \hat{n}_i^I(r)\hat{n}_i^I(r') - \frac{1}{2}\sum_{i<j} P_{ij}\hat{n}_i^I(r)\hat{n}_j^I(r')\right)}{|\vec{r} - \vec{r}'|}$$

$$+ \frac{4}{3}g^2 \sum_{I \neq J} \frac{N^I N^J}{2} \int\int d^3r \, d^3r'$$

$$\times \frac{\left(\sum_i P_{ii} \hat{n}_i^I(r)\hat{n}_i^J(r') - \frac{1}{2}\sum_{i<j} P_{ij}\hat{n}_i^I(r)\hat{n}_j^J(r')\right)}{|\vec{r} - \vec{r}'|}$$

$$+ \frac{4}{3}g^2 \sum_I N^I \frac{g^I (g^I - 1)}{2(g^I)^2} \int\int d^3r \, d^3r'$$

$$\times \frac{\left(\sum_i P_{ii} \hat{n}_i^I(r)\hat{n}_i^I(r') - \frac{1}{2}\sum_{i<j} P_{ij}\hat{n}_i^I(r)\hat{n}_j^I(r')\right)}{|\vec{r} - \vec{r}'|}$$

$$+ \frac{4}{3}g^2 \sum_I \frac{\bar{N}^I (\bar{N}^I - 1)}{2} \int\int d^3r \, d^3r'$$

$$\times \frac{\left(\sum_i \bar{P}_{ii} \hat{\bar{n}}_i^I(r)\hat{\bar{n}}_i^I(r') - \frac{1}{2}\sum_{i<j} \bar{P}_{ij}\hat{\bar{n}}_i^I(r)\hat{\bar{n}}_j^I(r')\right)}{|\vec{r} - \vec{r}'|}$$

$$+ \frac{4}{3}g^2 \sum_{I \neq J} \frac{\bar{N}^I \bar{N}^J}{2} \int\int d^3r \, d^3r'$$

$$\times \frac{\left(\sum_i \bar{P}_{ii} \hat{\bar{n}}_i^I(r)\hat{\bar{n}}_i^J(r') - \frac{1}{2}\sum_{i<j} \bar{P}_{ij}\hat{\bar{n}}_i^I(r)\hat{\bar{n}}_j^J(r')\right)}{|\vec{r} - \vec{r}'|}$$

$$+ \frac{4}{3}g^2 \sum_I \bar{N}^I \frac{\bar{g}^I(\bar{g}^I - 1)}{2(\bar{g}^I)^2} \int \int d^3r \, d^3r'$$

$$\times \frac{\left(\sum_i \bar{P}_{ii} \hat{\bar{n}}_i^I(r) \hat{\bar{n}}_i^I(r') - \frac{1}{2} \sum_{i<j} \bar{\bar{P}}_{ij} \hat{\bar{n}}_i^I(r) \hat{\bar{n}}_j^I(r') \right)}{|\vec{r} - \vec{r}'|}$$

$$- \frac{4}{3}g^2 \sum_{I,J} \bar{N}^I N^J \int \int d^3r \, d^3r'$$

$$\times \frac{\left(\sum_i \bar{P}_{ii} \hat{\bar{n}}_i^I(r) \hat{n}_i^J(r') - \frac{1}{2} \sum_{i<j} \bar{P}_{ij} \hat{\bar{n}}_i^I(r) \hat{n}_j^J(r') \right)}{|\vec{r} - \vec{r}'|}. \tag{2.26}$$

In Eq. (2.26) $\hat{n}_i^I$ are number densities, which are normalized to one when integrated over space [21]. Note that the degeneracy factors are already contained in the expression for number densities. So, we divided third and sixth terms by square of degeneracy factors.

From Table 2.5, we can see that the probability of interaction type CCA is twice the CCR type. Also, interaction probability AAA is twice the AAR type. This simple finding amazingly removes six out of seven terms from the interaction energy, leaving us with the last term from (2.26). Thus we conclude, for mesons on average, quarks only interact with antiquarks. This cancellation makes our analytical solutions easier to achieve.

The total energy E can now be written as the sum of kinetic and potential energies such that

$$E = \sum_{i,I} \int^{r_{max}} d^3r \frac{\left(6\pi^2 \hbar^3 N^I \hat{n}_i^I(r)\right)^{5/3}}{20\pi^2 \hbar^3 m^I (g^I)^{2/3}} + \sum_{i,I} \int^{r_{max}} d^3r \frac{\left(6\pi^2 \hbar^3 \bar{N}^I \hat{\bar{n}}_i^I(r)\right)^{5/3}}{20\pi^2 \hbar^3 \bar{m}^I (\bar{g}^I)^{2/3}}$$

$$+ \frac{4}{3}g^2 \sum_{I,J} \bar{N}^I N^J \int \int d^3r \, d^3r'$$

$$\times \frac{\left(\sum_i \bar{P}_{ii} \hat{\bar{n}}_i^I(r) \hat{n}_i^J(r') - \frac{1}{2} \sum_{i<j} \bar{P}_{ij} \hat{\bar{n}}_i^I(r) \hat{n}_j^J(r') \right)}{|\vec{r} - \vec{r}'|}. \tag{2.27}$$

Now, we switch back to normalization n_i^I and $\bar{n}_i^I$ and assume equal Fermi color momenta, $n^I \equiv n_1^I \equiv n_2^I \equiv n_3^I$ for each I. The same assumption is made for antiparticles. This gives

$$E = \sum_I \int^{r_{max}} d^3r \frac{3\left(6\pi^2 \hbar^3 N^I n^I(r)\right)^{5/3}}{20\pi^2 \hbar^3 m^I (g^I)^{2/3}} + \sum_I \int^{r_{max}} d^3r \frac{3\left(6\pi^2 \hbar^3 \bar{N}^I \bar{n}^I(r)\right)^{5/3}}{20\pi^2 \hbar^3 \bar{m}^I (\bar{g}^I)^{2/3}}$$

$$- \frac{9 \times 4/3 g^2}{2\eta - 1} \sum_{I,J} \int \int \frac{d^3r d^3r'}{|\vec{r} - \vec{r}'|} \bar{n}^I(r) n^J(r'). \tag{2.28}$$

Equations (2.20) and (2.21) can now be averaged over colors as

$$\int d^3r\, n^I(r) = N^I g^I/3,$$ (2.29)

and

$$\int d^3r\, \bar{n}^I(r) = \bar{N}^I \bar{g}^I/3.$$ (2.30)

We will use this pair of equations to set up the normalization conditions.

2.5   Thomas-Fermi Quark Equations

We can now introduce Lagrange's undetermined multipliers λ^I and $\bar{\lambda}^I$ associated with constraints in Eqs. (2.29) and (2.30) and add them to the expression for energy. This gives

$$E = \sum_{i,I} \int^{r_{max}} d^3r \frac{3\left(6\pi^2\hbar^3 N^I n^I(r)\right)^{5/3}}{20\pi^2\hbar^3 m^I} + \sum_{i,I} \int^{r_{max}} d^3r \frac{3\left(6\pi^2\hbar^3 \bar{N}^I \bar{n}^I(r)\right)^{5/3}}{20\pi^2\hbar^3 \bar{m}^I}$$

$$- \frac{9 \times 4/3 g^2}{2\eta - 1} \sum_{I,J} \int\int \frac{d^3r d^3r'}{|\vec{r} - \vec{r}'|} \bar{n}^I(r) n^J(r')$$

$$+ \sum_I \lambda^I \left(3\int d^3r n^I(r) - N^I g^I\right) + \sum_I \bar{\lambda}^I \left(3\int d^3r \bar{n}^I(r) - \bar{N}^I \bar{g}^I\right).$$ (2.31)

The purpose of adding these terms involving Lagrange multipliers is to allow a minimization of the total energy while keeping particle number constant. The density variations $\delta n^I(r)$ and $\delta n^I(r)$ give

$$\frac{(6\pi^2\hbar^3)^{5/3}}{\pi^2\hbar^3}\left[\frac{1}{4m^I}\left(\frac{n^I(r)}{g^I}\right)^{2/3}\right] = -3\lambda^I + \frac{9\times\frac{4}{3}g^2}{(2\eta-1)}\sum_I \int \frac{d^3r'}{|\vec{r}-\vec{r}'|}\bar{n}^I(r'),$$ (2.32)

$$\frac{(6\pi^2\hbar^3)^{5/3}}{\pi^2\hbar^3}\left[\frac{1}{4\bar{m}^I}\left(\frac{\bar{n}^I(r)}{g^I}\right)^{2/3}\right] = -3\bar{\lambda}^I + \frac{9\times\frac{4}{3}g^2}{(2\eta-1)}\sum_I \int \frac{d^3r'}{|\vec{r}-\vec{r}'|}n^I(r').$$ (2.33)

Assuming spherical symmetry, the TF spatial functions $f^I(r)$ and $\bar{f}^I(r)$ are defined such that

$$f^I(r) \equiv \frac{ra}{(8\alpha_s/3)} \left(\frac{6\pi^2 n^I(r)}{g^I} \right)^{2/3}, \tag{2.34}$$

$$\bar{f}^I(r) \equiv \frac{ra}{(8\alpha_s/3)} \left(\frac{6\pi^2 \bar{n}^I(r)}{\bar{g}^I} \right)^{2/3}, \tag{2.35}$$

where

$$a \equiv \frac{\hbar}{m^1 c} \tag{2.36}$$

gives the scale, where m^1 is the mass of lightest quark, and

$$\alpha_s = \frac{g^2}{\hbar c} \tag{2.37}$$

is the strong coupling constant. Equation (2.33) after using Eqs. (2.34)–(2.37) can now be written as

$$
\begin{aligned}
\frac{3m^1}{m^I} \frac{4}{3} g^2 \frac{\bar{f}^I(r)}{r} = & -3\bar{\lambda}^I + \frac{6 \cdot \frac{4}{3} g^2}{(2\eta - 1)\pi} \left(\frac{2 \times \frac{4}{3}\alpha_s}{a} \right)^{3/2} \\
& \times \sum_I g^I \left[\frac{1}{r} \int_0^r dr' r'^2 \left(\frac{f^I(r')}{r'} \right)^{3/2} \right. \\
& \left. + \int_0^{r_m} dr' r' \left(\frac{f^I(r')}{r'} \right)^{3/2} \right].
\end{aligned}
\tag{2.38}
$$

Here we have used the integral

$$\int^{r_{\max}} d^3r' \frac{n^I(r')}{|\vec{r} - \vec{r}'|} = 4\pi \left[\int_0^r dr' r'^2 \frac{n^I(r')}{r} + \int_r^{r_{\max}} dr' r'^2 \frac{n^I(r')}{r'} \right] \tag{2.39}$$

because of spherical symmetry. Equation (2.38) can be further simplified into

$$
\begin{aligned}
\alpha^I \bar{f}^I(r) = & \frac{-\bar{\lambda}^I r}{\frac{4}{3} g^2} + \frac{2}{(2\eta - 1)\pi} \left(\frac{2 \times \frac{4}{3}\alpha_s}{a} \right)^{3/2} \\
& \times \sum_I g^I \left[\int_0^r dr' r'^2 \left(\frac{f^I(r')}{r'} \right)^{3/2} + r \int_0^{r_m} dr' r' \left(\frac{f^I(r')}{r'} \right)^{3/2} \right],
\end{aligned}
\tag{2.40}
$$

where

$$\alpha^I = \frac{m^1}{m^I} \tag{2.41}$$

is the ratio of the lightest quark to the Ith quark.

Let us introduce the dimensionless parameter x such that $r = Rx$, where

$$R = \frac{a}{(8\alpha_s/3)} \left(\frac{3\pi\eta}{2}\right)^{2/3}. \tag{2.42}$$

Assuming spherical symmetry, Eq. (2.40) can now be written as

$$\alpha^I \bar{f}^I(x) = \frac{-\bar{\lambda}^I Rx}{\frac{4}{3}g^2} + \frac{3\eta}{2\eta - 1} \sum_I g^I$$

$$\times \left[\int_0^x dx'\, x'^2 \left(\frac{f^I(x')}{x'}\right)^{3/2} + x \int_x^{x_m} dx'\, x' \left(\frac{f^I(x')}{x'}\right)^{3/2} \right]. \tag{2.43}$$

Similarly, starting from Eq. (2.32) we can obtain another TF integral equation:

$$\alpha^I f^I(x) = \frac{-\lambda^I Rx}{\frac{4}{3}g^2} + \frac{3\eta}{2\eta - 1} \sum_I \bar{g}^I$$

$$\times \left[\int_0^x dx'\, x'^2 \left(\frac{\bar{f}^I(x')}{x'}\right)^{3/2} + x \int_x^{x_m} dx'\, x' \left(\frac{\bar{f}^I(x')}{x'}\right)^{3/2} \right]. \tag{2.44}$$

Taking first derivatives of Eqs. (2.43) and (2.44), we have

$$\alpha^I \frac{df^I(x)}{dx} = \frac{-\lambda^I R}{\frac{4}{3}g^2} + \frac{3\eta}{(2\eta - 1)} \left[\sum_I \bar{g}^I \int_x^{x_m} dx'\, x' \left(\frac{\bar{f}^I(x')}{x'}\right)^{3/2} \right], \tag{2.45}$$

and

$$\alpha^I \frac{d\bar{f}^I(x)}{dx} = \frac{-\bar{\lambda}^I R}{\frac{4}{3}g^2} + \frac{3\eta}{(2\eta - 1)} \left[\sum_I g^I \int_x^{x_{max}} dx'\, x' \left(\frac{f^I(x')}{x'}\right)^{3/2} \right]. \tag{2.46}$$

Similarly, second derivatives of Eqs. (2.43) and (2.44) yield the TF equations for mesonic matter. They are

$$\alpha^I \frac{d^2 f^I(x)}{dx^2} = -\frac{3\eta}{(2\eta - 1)} \frac{1}{\sqrt{x}} \sum_I \bar{g}^I \bar{f}^I(x)^{3/2}, \tag{2.47}$$

and

$$\alpha^I \frac{d^2 \bar{f}^I(x)}{dx^2} = -\frac{3\eta}{(2\eta - 1)} \frac{1}{\sqrt{x}} \sum_I g^I f^I(x)^{3/2}.$$ (2.48)

Equations (2.47) and (2.48) are the differential form of the TF quark equations in the case of mesons. The interchangeability of these equations shows the TF equations are invariant with respect to particle and antiparticle. As was mentioned before, it also shows that quarks interact only with antiquarks in mesonic matter; quark/quark and antiquark/antiquark interactions sum to zero in the TF model. When there is an explicit particle/antiparticle symmetry, we assume $f = \bar{f}$ to reduce the TF differential equations into one incredibly simple form:

$$\alpha^I \frac{d^2 f^I(x)}{dx^2} = -\frac{3\eta}{(2\eta - 1)} \frac{1}{\sqrt{x}} \sum_I g^I f^I(x)^{3/2}.$$ (2.49)

We use this equation to form system energies and equations for charmonium–bottomonium family and Z-meson family systems. However, for D-meson families, the explicit asymmetry in the masses of the particle and antiparticle will require us to take a different approach.

2.6 Energy Equations in Terms of Dimensionless Radius

The expressions for kinetic and potential energies now need to be given in terms of the dimensionless distance in order for us to be able to obtain analytical solutions. In addition, the volume energy term that produces an external pressure on the system needs to be appropriately introduced.

To obtain the kinetic and potential energy expressions, we begin from Eq. (2.28) and apply Eqs. (2.34), (2.35), and (2.42). This gives the kinetic energy as

$$T = \sum_I \frac{12}{5\pi} \left(\frac{3\pi\eta}{2}\right)^{1/3} \frac{\frac{4}{3}g^2 \cdot \frac{4}{3}\alpha_s}{a} \alpha_I g^I \int_0^{x_I} dx \frac{\left(f^I(x)\right)^{5/2}}{\sqrt{x}}$$

$$+ \sum_I \frac{12}{5\pi} \left(\frac{3\pi\eta}{2}\right)^{1/3} \frac{\frac{4}{3}g^2 \cdot \frac{4}{3}\alpha_s}{a} \alpha_I \bar{g}^I \int_0^{x_I} dx \frac{\left(\bar{f}^I(x)\right)^{5/2}}{\sqrt{x}}.$$ (2.50)

The potential energy

$$U = -\frac{9 \cdot \frac{4}{3}g^2}{(2\eta - 1)} \sum_{I,J} \bar{g}^I g^J \int^{r_I} \int^{r_J} d^3r \, d^3r' \frac{n^I(r)n^J(r')}{|\vec{r} - \vec{r}'|}$$ (2.51)

becomes

$$U = -\frac{9 \cdot \frac{4}{3} g^2}{(2\eta - 1)} \frac{\eta^2}{R} \sum_{I,J} \bar{g}^I g^J \Big[\int_0^{x_I} dx \frac{\left(\bar{f}^I(x)\right)^{3/2}}{\sqrt{x}} \int_0^x dx' \sqrt{x'} \left(f^J(x')\right)^{3/2}$$

$$+ \int_0^{x_I} dx \left(\bar{f}^I(x)\right)^{3/2} \sqrt{x} \int_x^{x_J} dx' \frac{\left(f^J(x')\right)^{3/2}}{\sqrt{x'}} \Big].$$

(2.52)

The volume energy (E_v) term gives the inward pressure which keeps quarks within a boundary. We assume that [17]

$$E_v = \frac{4}{3}\pi R^3 x_{\max}^3 B,$$

(2.53)

where B is the bag constant. Now that we have T, U, and E_v terms, we can find the total energy of a desired multi-quark state. The total energy of such a state is simply given by

$$E = T + V + E_v + \eta \cdot m_q + \eta \cdot \bar{m}_q,$$

(2.54)

where m_q and $\bar{m}_q$ are the mass of the quark and antiquark, respectively.

In Sect. 2.8 we will fit model parameters for a given set of mesons, including m_q and $\bar{m}_q$. However, to assess the stability of multi-quark mesons we omit the trivial mass part and will examine the energy as a function of quark content.

2.7   Application of Model

In this section we will now apply our model and obtain energy expressions for different families of quarks. In all the cases that follow charm can be read as bottom also.

2.7.1   Charmonium Family: Case 1

This type of family contains quarks and antiquarks of equal mass. The system of quarks that lies inside this family can be represented as $Q\bar{Q}$, $Q\bar{Q}Q\bar{Q}$, $Q\bar{Q}Q\bar{Q}Q\bar{Q}$, and so on. We have investigated the case, where Q is a charm quark or bottom quark but not both in the same system. We call it the charmonium family. Charmonium and all the multi-quark families of charmonium consist of charm and anticharm (or bottom and anti-bottom) only. Therefore, I takes a single value and hence is dropped. g_0 refers to degeneracy and, for this application, can have the value of one or two to represent spin. This can give spin splittings, but as we pointed above we have not yet included explicit spin–spin interactions in the model, like we have done in our baryon model [24]. Equation (2.49) becomes

$$\frac{d^2 f(x)}{dx^2} = -\frac{3\eta}{(2\eta - 1)} \cdot g_0 \cdot \frac{1}{\sqrt{x}} f(x)^{3/2}, \tag{2.55}$$

where

$$g_0 \times N^I = \eta. \tag{2.56}$$

We choose the normalization equation to be

$$\int_0^{x_{\max}} dx \sqrt{x} f(x)^{3/2} = \frac{N^I}{3\eta}. \tag{2.57}$$

Expressing the normalization equation in terms of boundary conditions, we get

$$\left(x\frac{df}{dx} - f \right) |_{x_{\max}} = -\frac{\eta}{2\eta - 1}. \tag{2.58}$$

With these modified boundary conditions, we can derive the expression for kinetic and potential energies. For the kinetic energy we can start with Eq. (2.50) and assume a single flavor. In this case, both quarks and antiquarks contribute equally to the kinetic energy, which gives

$$T = 2 \cdot \frac{12}{5\pi} \left(\frac{3\pi\eta}{2} \right)^{1/3} \frac{\frac{4}{3}g^2 \cdot \frac{4}{3}\alpha_s}{a} g_0 \int_0^{x_{\max}} dx \frac{(f(x))^{5/2}}{\sqrt{x}}. \tag{2.59}$$

The integral may be done using the TF differential equation, and results in

$$T = \frac{24}{5\pi} \left(\frac{3\pi\eta}{2} \right)^{1/3} \frac{\frac{4}{3}g^2 \cdot \frac{4}{3}\alpha_s}{a} \left[-\frac{5}{21} \frac{df(x)}{dx} |_{x_{\max}} + \frac{4}{7}\sqrt{x_{\max}} \left(f(x_{\max}) \right)^{5/2} g_0 \right]. \tag{2.60}$$

Thus, the kinetic energy depends only on the derivative and value of the TF function at the boundary. When there is a single flavor as in the case of the charmonium family, and symmetry of particle and antiparticle occurs, the potential energy Eq. (2.52) can be written as

$$U = -\frac{9 \cdot \frac{4}{3}g^2}{(2\eta - 1)} \frac{\eta^2}{R} \left(g^I \right)^2 \left[\int_0^{x_{\max}} dx \frac{(f(x))^{3/2}}{\sqrt{x}} \int_0^x dx' \sqrt{x'} \left(f(x') \right)^{3/2} \right.$$
$$\left. + \int_0^{x_{\max}} dx (f(x))^{3/2} \sqrt{x} \int_x^{x_{\max}} dx' \frac{(f(x'))^{3/2}}{\sqrt{x'}} \right]. \tag{2.61}$$

This can be further simplified to

$$U = \frac{4}{\pi} \left(\frac{3\pi\eta}{2} \right)^{1/3} \frac{\frac{4}{3}g^2 \cdot \frac{4}{3}\alpha_s}{a} \times \left[\frac{4}{7} \frac{df(x)}{dx} \Big|_{x_{\max}} - \frac{4}{7} \sqrt{x_{\max}} \, (f(x_{\max}))^{5/2} \, g_0 \right],$$

which like the expression for T depends on the derivative and values of the TF function at the boundary.

2.7.2 Z-Meson Family: Case 2

The constituents of Z-mesons are charm (c) (or bottom(b)), anticharm ($\bar{c}$) (or anti-bottom ($\bar{b}$)), light (u or d), and antilight quarks. The particles in this family can be represented as $Q\bar{Q}q\bar{q}$, $Q\bar{Q}q\bar{q}Q\bar{Q}q\bar{q}$, $Q\bar{Q}q\bar{q}Q\bar{Q}q\bar{q}Q\bar{Q}q\bar{q}$, and so on, where Q represents a heavy quark and q is a light quark. We treat the mass of the up and down quarks as the same. This means the Z-meson and all multi-quark families of Z-mesons have a total quark mass equal to the antiquark mass. As before, we set $f = \bar{f}$ in the TF equations and obtain Eq. (2.49) with $I = 1$ or 2. Let $f^1(x)$ be the TF function of the light quark and $f^2(x)$ be the TF function of the heavier quark. For N_1 quarks with degeneracy factor g_1 and N_2 quarks with degeneracy g_2, we have

$$g_1 N_1 + g_2 N_2 = \eta. \tag{2.62}$$

In our application g_1 can have values one, two, or four, whereas g_2 can have value of one or two only. We assume a linear relation exists between $f^1(x)$ and $f^2(x)$ in the region $0 < x < x_2$, and that $f^2(x)$ vanishes for a dimensionless distance greater than x_2, i.e.,

$$f^1(x) = k f^2(x) \quad \text{for} \quad 0 \le x \le x_2,$$
$$f^2(x) = 0 \qquad \text{for} \quad x_2 \le x \le x_1. \tag{2.63}$$

Equation (2.49) can now be written as two set of equations:

$$\alpha^1 \frac{d^2 f^1(x)}{dx^2} = -\frac{3\eta}{(2\eta - 1)} \frac{1}{\sqrt{x}} \left(g_1 \left(f^1(x) \right)^{3/2} + g_2 \left(f^2(x) \right)^{3/2} \right), \tag{2.64}$$

$$\alpha^2 \frac{d^2 f^2(x)}{dx^2} = -\frac{3\eta}{(2\eta - 1)} \frac{1}{\sqrt{x}} \left(g_1 \left(f^1(x) \right)^{3/2} + g_2 \left(f^2(x) \right)^{3/2} \right). \tag{2.65}$$

To make Eqs. (2.64) and (2.65) consistent, we need

$$k = \frac{\alpha^2}{\alpha^1}, \tag{2.66}$$

which is just the inverse ratio of the given masses from (2.41). The similar step in the case of baryons gives a much more complicated consistency condition [24]. The normalization conditions are

$$\int_0^{x_2} x^{1/2} \left(f^2(x) \right)^{3/2} dx = \frac{N_2}{3\eta}, \tag{2.67}$$

and

$$\int_0^{x_1} x^{1/2} \left(f^1(x) \right)^{3/2} dx = \frac{N_1}{3\eta}. \tag{2.68}$$

In region $0 < x < x_2$,

$$\frac{d^2 f^1(x)}{dx^2} = Q_1 \frac{\left(f^1(x) \right)^{3/2}}{\sqrt{x}}, \tag{2.69}$$

where

$$Q_1 = -\frac{3\eta}{(2\eta - 1)\alpha_1} \left(g_1 + \frac{g_2}{k^{3/2}} \right). \tag{2.70}$$

In region $x_2 < x < x_1$,

$$\frac{d^2 f^1(x)}{dx^2} = Q_2 \frac{\left(f^1(x) \right)^{3/2}}{\sqrt{x}}, \tag{2.71}$$

where

$$Q_2 = -\frac{3\eta}{(2\eta - 1)\alpha_1} g_1. \tag{2.72}$$

With the equations above we can express normalization conditions in the form of boundary conditions:

$$\left(x \frac{df^2(x)}{dx} - f^2(x) \right)\bigg|_{x_2} = -\frac{N_2 Q_1 k^{3/2}}{3\eta}, \tag{2.73}$$

$$\left(x \frac{df^1(x)}{dx} - f^1(x) \right)\bigg|_{x_1} = -\frac{\eta}{(2\eta - 1)\alpha_1}. \tag{2.74}$$

The energies are derived as for Case 1. For g_1 flavors with N_1 particles and g_2 flavors with N_2 particles we have

$$T = 2 \cdot \frac{12}{5\pi} \left(\frac{3\pi\eta}{2} \right)^{1/3} \frac{\frac{4}{3}g^2 \cdot \frac{4}{3}\alpha_s}{a}$$

$$\times \left[g_1\alpha_1 \int_0^{x_1} \frac{\left(f^1(x)\right)^{5/2}}{\sqrt{x}} dx + g_2\alpha_2 \int_0^{x_2} \frac{\left(f^2(x)\right)^{5/2}}{\sqrt{x}} dx \right].$$

Using the TF function differential equations, the consistency condition for k and boundary conditions allows one to relate the integrals to TF function values and derivatives on the surfaces:

$$T = 2 \cdot \frac{12}{5\pi} \left(\frac{3\pi\eta}{2} \right)^{1/3} \frac{\frac{4}{3}g^2 \cdot \frac{4}{3}\alpha_s}{a} \left[-\frac{5}{21}\alpha_1 \frac{df^1(x)}{dx}|_{x_1} \right. \tag{2.75}$$

$$\left. + \frac{4}{7}\sqrt{x_1} \left(f^1(x_1)\right)^{5/2} g_1\alpha_1 + \frac{4}{7}\sqrt{x_2} \left(f^2(x_2)\right)^{5/2} g_2\alpha_2 \right].$$

The expression for potential energy in the same case becomes

$$U = -\frac{9 \cdot \frac{4}{3}g^2}{(2\eta - 1)} \frac{\eta^2}{R} \sum_{I,J} g^I g^J \left[\int_0^{x_I} dx \frac{\left(f^I(x)\right)^{3/2}}{\sqrt{x}} \int_0^x dx' \sqrt{x'} \left(f^J(x')\right)^{3/2} \right.$$

$$\tag{2.76}$$

$$\left. + \int_0^{x_I} dx \left(f^I(x)\right)^{3/2} \sqrt{x} \int_x^{x_J} dx' \frac{\left(f^J(x')\right)^{3/2}}{\sqrt{x'}} \right],$$

which can be placed into the form

$$U = -\frac{9 \cdot \frac{4}{3}g^2}{(2\eta - 1)} \frac{\eta^2}{R} \left[g_1^2 K_1 + g_2^2 K_2 + g_1 g_2 K_{12} + g_1 g_2 K_{21} \right], \tag{2.77}$$

where

$$K_1 \equiv \int_0^{x_1} dx \frac{\left(f^1(x)\right)^{3/2}}{\sqrt{x}} \int_0^x dx' \sqrt{x'} \left(f^1(x')\right)^{3/2} \tag{2.78}$$

$$+ \int_0^{x_1} dx \left(f^1(x)\right)^{3/2} \sqrt{x} \int_x^{x_1} dx' \frac{\left(f^1(x')\right)^{3/2}}{\sqrt{x'}},$$

$$K_2 \equiv \int_0^{x_2} dx \frac{\left(f^2(x)\right)^{3/2}}{\sqrt{x}} \int_0^x dx' \sqrt{x'} \left(f^2(x')\right)^{3/2} \tag{2.79}$$

$$+ \int_0^{x_2} dx \left(f^2(x) \right)^{3/2} \sqrt{x} \int_x^{x_2} dx' \frac{\left(f^2(x') \right)^{3/2}}{\sqrt{x'}},$$

$$K_{12} \equiv \int_0^{x_1} dx \frac{\left(f^1(x) \right)^{3/2}}{\sqrt{x}} \int_0^x dx' \sqrt{x'} \left(f^2(x') \right)^{3/2} \tag{2.80}$$

$$+ \int_0^{x_1} dx \left(f^1(x) \right)^{3/2} \sqrt{x} \int_x^{x_2} dx' \frac{\left(f^2(x') \right)^{3/2}}{\sqrt{x'}},$$

and

$$K_{21} \equiv \int_0^{x_2} dx \frac{\left(f^2(x) \right)^{3/2}}{\sqrt{x}} \int_0^x dx' \sqrt{x'} \left(f^1(x') \right)^{3/2} \tag{2.81}$$

$$+ \int_0^{x_2} dx \left(f^2(x) \right)^{3/2} \sqrt{x} \int_x^{x_1} dx' \frac{\left(f^1(x') \right)^{3/2}}{\sqrt{x'}}.$$

One can show that the K_{12} integral is equivalent to the K_{21} integral.

At this point we would like to take some time to explain the reductions of these expressions. We will see some interesting cancellations in the analytical solutions. The expressions for K_1, K_2, and K_{12} after integrations and reductions can be written as

$$
\begin{aligned}
K_1 = &\int_0^{x_2} dx \frac{\left(f^1(x) \right)^{5/2}}{\sqrt{x}} \left(-\frac{1}{Q_1} \right) \\
&+ \int_0^{x_2} dx \sqrt{x} \left(f^1(x) \right)^{3/2} \left(\frac{1}{Q_1} \left. \frac{df^1(x)}{dx} \right|_{x_2} \right. \\
&+ \frac{1}{Q_2} \left. \frac{df^1(x)}{dx} \right|_{x_1} - \frac{1}{Q_2} \left. \frac{df^1(x)}{dx} \right|_{x_2} \Bigg) \\
&+ \int_{x_2}^{x_1} dx \frac{\left(f^1(x) \right)^{3/2}}{\sqrt{x}} \left(\frac{N_2 k^{3/2}}{\eta} \left(1 - \frac{Q_1}{Q_2} \right) \right) \\
&+ \int_{x_2}^{x_1} dx \frac{\left(f^1(x) \right)^{5/2}}{\sqrt{x}} \left(-\frac{1}{Q_2} \right) \\
&+ \int_{x_2}^{x_1} dx \left(f^1(x) \right)^{3/2} \sqrt{x} \left(\frac{1}{Q_2} \left. \frac{df^1(x)}{dx} \right|_{x_1} \right),
\end{aligned}
\tag{2.82}
$$

$$K_2 = \int_0^{x_2} dx \frac{\left(f^2(x)\right)^{5/2}}{\sqrt{x}} \left(-\frac{1}{Q_1 k^{1/2}}\right) \tag{2.83}$$

$$+ \int_0^{x_2} dx \left(f^2(x)\right)^{3/2} \sqrt{x} \left(\frac{1}{Q_1 k^{1/2}} \left.\frac{df^2(x)}{dx}\right|_{x_2}\right),$$

and

$$K_{12} = \int_0^{x_2} dx \frac{\left(f^2(x)\right)^{5/2}}{\sqrt{x}} \left(-\frac{k}{Q_1}\right) \tag{2.84}$$

$$+ \int_0^{x_2} dx \left(f^2(x)\right)^{5/2} \sqrt{x} \left(\frac{k}{Q_1} \left.\frac{df^2(x)}{dx}\right|_{x_2}\right)$$

$$+ \int_{x_2}^{x_1} \frac{\left(f^1(x)\right)^{3/2}}{\sqrt{x}} \left(\frac{N_2}{\eta}\right).$$

Similarly, the potential energy now simplifies to

$$U = -\frac{9 \cdot \frac{4}{3} g^2}{(2\eta - 1)} \frac{\eta^2}{R} \times \left[\int_0^{x_2} dx \frac{\left(f^2(x)\right)^{5/2}}{\sqrt{x}} \left[\frac{-k^{5/2}}{Q_1} g_1^2 - \frac{1}{Q_1 k^{1/2}} g_2^2 - \frac{k}{Q_1} 2 g_1 g_2\right]\right.$$
$$\tag{2.85}$$

$$+ \int_0^{x_2} dx \left(f^2(x)\right)^{3/2} \sqrt{x} \times \left[g_1^2 \left(\frac{k^{5/2}}{Q_1} \left.\frac{df^2(x)}{dx}\right|_{x_2} + \frac{k^{3/2}}{Q_2} \left.\frac{df^1(x)}{dx}\right|_{x_1}\right.\right.$$

$$\left.- \frac{k^{5/2}}{Q_2} \left.\frac{df^2(x)}{dx}\right|_{x_2}\right) + g_2^2 \frac{1}{Q_1 k^{1/2}} \left.\frac{df^2(x)}{dx}\right|_{x_2} + 2 g_1 g_2 \frac{k}{Q_1} \left.\frac{df^2(x)}{dx}\right|_{x_2}\right]$$

$$+ \int_{x_2}^{x_1} dx \frac{\left(f^1(x)\right)^{3/2}}{\sqrt{x}} \left[\frac{N_2 k^{3/2}}{\eta} \left(1 - \frac{Q_1}{Q_2}\right) g_1^2 + \frac{N_2}{\eta} 2 g_1 g_2\right]$$

$$+ \int_{x_2}^{x_1} dx (f^1(x))^{3/2} \sqrt{x} \left[\frac{1}{Q_2} \left.\frac{df^1(x)}{dx}\right|_{x_1} g_1^2\right]$$

$$+ \int_{x_2}^{x_1} \frac{\left(f^1(x)\right)^{5/2}}{\sqrt{x}} \left[-\frac{1}{Q_2} g_1^2\right]\right],$$

which on further reduction yields

$$
U = -\frac{9 \cdot \frac{4}{3}g^2}{(2\eta - 1)} \frac{\eta^2}{R} \times \left[\frac{N_2}{\eta} \frac{df^2(x)}{dx} \bigg|_{x_2} \times \left[\frac{5}{7} \frac{k^{5/2}g_1^2}{Q_1} + \frac{5}{7} \frac{g_2^2}{Q_1 k^{1/2}} + \frac{5}{7} \frac{k}{Q_1} 2g_1g_2 \right. \right.
$$

(2.86)

$$
+ \frac{k^{5/2}g_1^2}{Q_1} - \frac{k^{5/2}g_1^2}{Q_2} + \frac{g_2^2}{Q_1 k^{1/2}} + \frac{k}{Q_1} 2g_1g_2 - \frac{k^{5/2}}{Q_2}\left(1 - \frac{Q_1}{Q_2}\right) g_1^2
$$

$$
\left. - \frac{k}{Q_2} 2g_1g_2 - \frac{5}{7Q_2} Q_1 k^{5/2} \frac{g_1^2}{Q_2} \right] + \frac{df^1(x)}{dx}
$$

$$
\times \left[\frac{N_2}{\eta} \frac{k^{3/2}g_1^2}{Q_2} + \frac{N_2 k^{3/2}}{\eta Q_2}\left(1 - \frac{Q_1}{Q_2}\right) g_1^2 + \frac{N_2}{\eta Q_2} 2g_1g_2 \right.
$$

$$
\left. \left. + \frac{g_1^2}{Q_2^2}\left(\frac{-3\eta}{(2\eta - 1)\alpha_1} - \frac{N_2 Q_1 k^{3/2}}{\eta}\right) - \frac{g_1^2}{Q_2^2}\left(\frac{5}{7}\frac{-3\eta}{(2\eta - 1)\alpha_1}\right) \right] \right.
$$

$$
+ \frac{4}{7}\sqrt{x_2}(f^2(x_2))^{5/2} \left[\frac{-k^{5/2}g_1^2}{Q_1} - \frac{g_2^2}{Q_1 k^{1/2}} - \frac{k}{Q_1} 2g_1g_2 + \frac{k^{5/2}g_1^2}{Q_2} \right]
$$

$$
\left. + \frac{4}{7}\sqrt{x_1}(f^1(x_1))^{5/2} \left[-\frac{g_1^2}{Q_2} \right] \right].
$$

Here we observe that coefficient of $\frac{N_2}{\eta}\frac{df^2(x)}{dx}|_{x_2}$ vanishes. All the above eleven terms actually cancel. Also all the other seemingly difficult integrals boil down to a simple equation, which gives

$$
U = -\frac{4}{\pi}\left(\frac{3\pi\eta}{2}\right)^{1/3} \frac{\frac{4}{3}g^2 \cdot \frac{4}{3}\alpha_s}{a}
$$

(2.87)

$$
\times \left[-\frac{4}{7}\alpha_1 \frac{df^1(x)}{dx}\bigg|_{x_1} + \frac{4}{7}\sqrt{x_1}\left(f^1(x_1)\right)^{5/2} g_1\alpha_1 + \frac{4}{7}\sqrt{x_2}\left(f^2(x_2)\right)^{5/2} g_2\alpha_2 \right].
$$

So, the interesting thing we observe in both the kinetic and potential energy expressions is that there is no dependency on the derivative of TF function at the inner boundary, unlike the baryon case.

2.7.3 D-Meson Family: Case 3

The D-meson and multi-quark families of D-mesons consist of charm (or bottom) quarks (or antiquarks) and light antiquarks (or light quarks). All the quarks in this family can be represented as $Q\bar{q}$, $Q\bar{q}Q\bar{q}$, $Q\bar{q}Q\bar{q}Q\bar{q}$, and so on. We will study the systems where the heavier mass is a quark and the lighter mass is an antiquark in each member of the D-meson family. By the symmetry inherent in the TF equations, this also covers the antiparticle state where the particle and antiparticle are interchanged. There is asymmetry in total mass of quarks and antiquarks, unlike the other cases we have studied. Also, because quarks interact only with antiquarks, the mathematics is different than earlier cases. We have two different TF functions for the charm and antilight quark, f and $\bar{f}$, respectively. We assume there is a universal TF function $\bar{f} = k_0 f$ in the region where the TF functions overlap. This means they are related linearly in the smaller region where the TF function of the charmed quark is nonzero. Outside of this, only the TF function of the light antiquark exists.

For the region $0 < x < x_2$, we have the differential equation

$$\frac{d^2 f(x)}{dx^2} = Q_0 \frac{(f(x))^{3/2}}{\sqrt{x}}, \tag{2.88}$$

where

$$Q_0 = -\frac{3\eta\, g_0}{(2\eta - 1)\cdot k_0}, \tag{2.89}$$

and

$$k_0 = \left(\frac{g_0\,\alpha}{\bar{g}_0\,\bar{\alpha}}\right)^{2/5} \tag{2.90}$$

is the consistency condition. α is given by Eq. (2.41), and in this case $\bar{\alpha} = 1$.

In region $x_2 < x < x_1$ with two TF equations, one function is zero and the other becomes

$$\frac{d^2 \bar{f}(x)}{dx^2} = 0,$$
$$\implies \bar{f}(x) = c \cdot x + d, \tag{2.91}$$

where

$$c = k_0 f'(x_2), \tag{2.92}$$

and

$$d = k_0 \left(f(x_2) - f'(x_2) \cdot x_2 \right). \tag{2.93}$$

Proceeding as before we obtained expressions for kinetic and potential energies. The energies are

$$T = \frac{12}{5\pi} \left(\frac{3\pi\eta}{2} \right)^{1/3} \frac{\frac{4}{3}g^2 \cdot \frac{4}{3}\alpha_s}{a} \tag{2.94}$$

$$\times \left[-\frac{10}{21}\alpha \frac{df(x)}{dx} \bigg|_{x_2} + \frac{8}{7}\sqrt{x_2}\,(f(x_2))^{5/2}\,g_0\alpha + \bar{g}_0 \int_{x_2}^{x_1} \frac{\left(\bar{f}(x) \right)^{5/2}}{\sqrt{x}}\,dx \right],$$

$$U = -\frac{4}{\pi} \left(\frac{3\pi\eta}{2} \right)^{1/3} \frac{\frac{4}{3}g^2 \cdot \frac{4}{3}\alpha_s}{a} \tag{2.95}$$

$$\times \left[-\frac{4}{7}\alpha \frac{df(x)}{dx} \bigg|_{x_2} \frac{\bar{g}_0}{g_0} + \frac{4}{7}\sqrt{x_2}\,(f(x_2))^{5/2}\,\bar{g}_0\alpha + \frac{\eta\,\bar{g}_0}{2\eta - 1} \int_{x_2}^{x_1} \frac{\left(\bar{f}(x) \right)^{3/2}}{\sqrt{x}}\,dx \right].$$

As a consistency check, if we assume light and heavy quarks to have equal mass, we have found that the potential and kinetic energies with $\eta = 1$ in Case 3 are equal to that of Case 1. Also, the non-degenerate Case 2 with $\eta = 2$ is the same as degenerate Case 1 with $\eta = 2$. We confirmed these both in our analytical and numerical results.

2.8 Method and Remarks

The phenomenological parameters we need for our model are the strong coupling constant α_s, the bag constant B, the charm and bottom quark masses, m_c and m_b, as well as the light quark mass, m_1. Previously, we used baryon phenomenology to obtain these parameters [26]. Also we did a fitting using mesonic states that involves only the charm quarks [27]. Here, we have included bottom quarks as well and obtained the fitted parameters. Since we do not yet include spin interactions in our model, we need to weight spin-split states to "remove" this interaction for our model fits.

Thus, we first fit the model expressions for the masses of charmonium states, bottomonium states, spin-weighted D-meson states, spin-weighted B-meson states, a spin-weighted combination of spin 1 likely tetraquark states involving charm quarks and spin 1 likely tetraquark states involving bottom quarks. In this way, we obtained the spin-weighted mass for these six states and then fitted the parameters above to obtain the minimized chi-square.

For the mass of charmonium and bottomonium we weighted masses of $1S$ states such that

$$\frac{1}{4}\left(\eta_c\,(1S)\right) + \frac{3}{4}\left(J/\Psi\,(1S)\right) = 3069\ \text{MeV} \tag{2.96}$$

$$\frac{1}{4}\left(\eta_b\,(1S)\right) + \frac{3}{4}\left(Y\,(1S)\right) = 9445\ \text{MeV}, \tag{2.97}$$

which we will refer to as Case 1-charm and Case 1-bottom mass.

Calculation of Case 2-charm and Case 2-bottom masses were a little trickier. To obtain Case 2-charm, we weighted spin 1 likely tetraquark states called the $Z_c(3900)$ and $X(4020)$. There are a number of charmonium-like exotic resonances which have been discovered in recent years [28]. The ground state of this set is the $\chi_{c1}(3872)$, with a mass of 3871.7 MeV, discovered by the Belle Collaboration in 2003 [6]. However, this state appears to be an isospin singlet, although the charge states might not have been seen yet [29]. On the other hand, the $Z_c(3900)$ resonance, with mass 3886.6 MeV, spin 1, and $C = -1$, is a triplet as we would expect from the hidden charm nonrelativistic tetraquark model. In addition, the fact that the $Z_c(3900)$ is above the $D^0\bar{D}^{*+}$ threshold makes it more likely that it is a true tetraquark rather than a molecular state of these same two particles. In the same way, the $X(4020)$ with mass 4024.1 MeV is a spin 1, $C = -1$ state which in this case is just above $D^{*0}\bar{D}^{*+}$ threshold. We will adopt these two particles as our spin-split charmed tetraquark states. Note that there are hidden bottom analogs of the $Z_c(3900)$ and $X(4020)$ states in the $Z_b(10610)$ and $Z_b(10650)$ states. We will use these to define Case 2 Z-mesons.

Case 2 calculations are now simply

$$\frac{1}{4} \times Z_c(3900) + \frac{3}{4} \times X(4020) = 3990\ \text{MeV}, \tag{2.98}$$

and

$$\frac{1}{4} \times Z_b(10610) + \frac{3}{4} \times Z_b(10650) = 10641\ \text{MeV} \tag{2.99}$$

which we will refer to as Case 2-charm and Case 2-bottom mass, respectively.

For the mass of D-meson we weighted spin zero particles

$$D\ (\text{spin zero average}) = \frac{1}{2}\left(D^+\right) + \frac{1}{2}\left(D^0\right) = 1866.5\ \text{MeV}, \tag{2.100}$$

Table 2.6 Comparison between masses used for fitting and those obtained after fitting

Fitted particle	Masses after fit(MeV)	Masses used for fit(MeV)
Case 1-charm	3049	3069
Case 2-charm	4015	3990
Case 3-charm	1960	1973
Case 1-bottom	9469	9445
Case 2-bottom	10,653	10,641
Case 3-bottom	5239	5313

with spin one D-mesons and obtained Case 3-charm. Case 3-charm equals

$$\frac{1}{4} D \text{ (spin zero average)} + \frac{3}{4} D \text{ (spin one average)} = 1973 \text{ MeV}. \tag{2.101}$$

Similarly for B-mesons, we obtained Case 3-bottom value as 5313 MeV.

Using *Mathematica*, we fitted parameters such that the mass chi-square was minimized using a grid search, obtaining $\sqrt{\chi^2} = 86.1$ MeV. We solved the differential equations using an iterative implementation of **NDSolve** in *Mathematica*. We obtained $\alpha_s = 0.346$, $B^{1/4} = 107.6$ MeV, charm quark mass $m_c = 1553$ MeV, and bottom quark mass $m_b = 4862$ MeV. Earlier paper uses a different parameter set than is used here. Our light quark mass, $m_1 = 306$ MeV, we take from our previous TF baryon fit [24]. Note that earlier paper containing some preliminary results is given in Refs. [26, 27]. Both proceedings papers also do not consider the bottom quark sector, as we do here. Table 2.6 gives us the difference between expected and obtained masses.

We will examine TF functions and energies for the three cases defined above. In nuclear physics, one examines the binding energy per nucleon in order to assess the stability of a given nucleus. We will do a similar investigation here. Thus, the important figure of merit in these evaluations is the total energy per quark pair, for if this increases as one adds more quarks, the family is unstable under decay to lower family members, whereas if it decreases, the family is stable. The static mass quark dependence of the mesons does not play a role in these considerations and so will be left off.

2.9 Results and Discussions

First, let us discuss the behavior of the density TF functions. The particle density function is proportional to $(f(x)/x)^{3/2}$ for all states. The charmonium and bottomonium function decreases with increase in distance and vanishes at a boundary

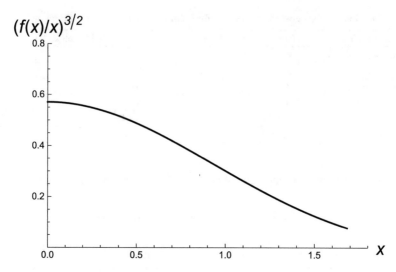

Fig. 2.1 TF density function of charmonium

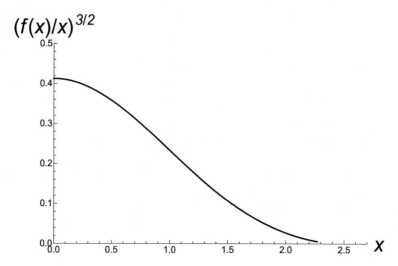

Fig. 2.2 TF density function of bottomonium

as seen in Figs. 2.1 and 2.2, where the dimensionless x variable is used. The density function for bottomonium is almost zero at the boundary.

The density function of Z-mesons has a long tail for the light quarks, while for charmed quarks the value is large and is concentrated near the origin, as seen in Figs. 2.3 and 2.4. This suggests an atomic-like structure with heavy charm, anticharm quarks at the center, while light quarks and antiquarks spread out like

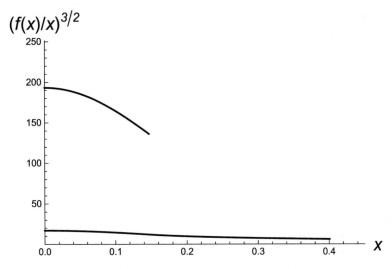

Fig. 2.3 TF density function of Case 2-mesons with charm quarks($c\bar{c}u\bar{u}$)

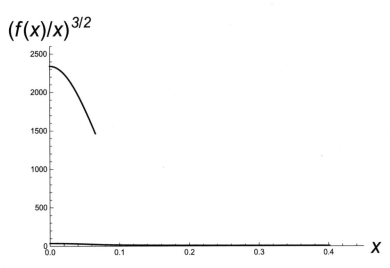

Fig. 2.4 Density function of Case 2-mesons with bottom quarks($b\bar{b}u\bar{u}$)

electrons. Figures 2.5 and 2.6 show an enlargement of the density function of the light quark TF function for the Z-meson. It drops down abruptly until it reaches the boundary of the heavy quark TF function, then inflects and decreases. In the case of D- and B-mesons, Figs. 2.7 and 2.8, respectively, the density function of light and heavy quarks is relatively closer. We can see that when charm is interchanged with bottom quarks density has a larger value for the heavier one as it should.

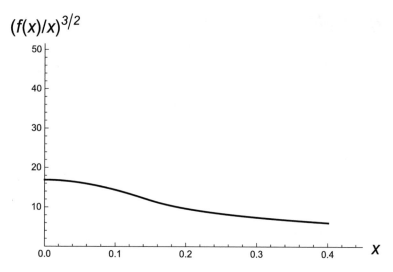

Fig. 2.5 Density function of light quarks for Case 2-mesons involving charm quarks($c\bar{c}u\bar{u}$)

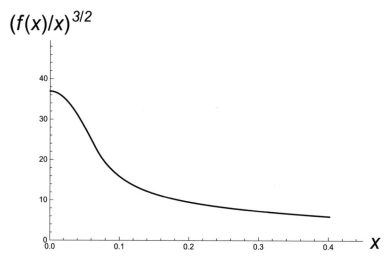

Fig. 2.6 Density function of light quarks for Case 2-mesons involving bottom quarks($b\bar{b}u\bar{u}$)

We increased the quark content and compared density functions of a family of multi-mesons in all three cases. We observed similar density functions for a given multi-meson family regardless of the quark content.

Figures 2.9 and 2.10 include 11 possibilities for the physical radius. The physical radius is plotted versus quark number and compared with a generic baryon with three degenerate light flavors. From Eq. (2.42) we can see that the radius is proportional to the product of $\eta^{2/3}$ and the dimensionless radius x. As the quark

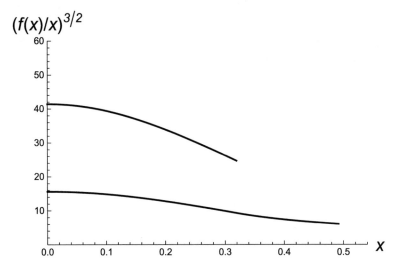

Fig. 2.7 Density of TF functions for the D-meson($c\bar{u}$)

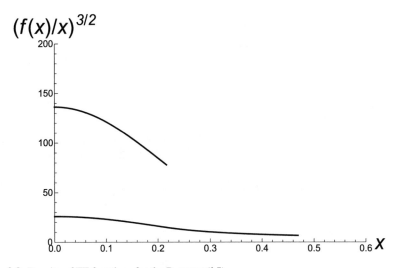

Fig. 2.8 Density of TF functions for the B-meson($b\bar{u}$)

content increases, the dimensionless radius, x, becomes smaller, whereas $\eta^{2/3}$ increases. In most cases, the result is an increase in radius with increasing quark content. We also observe that the curve of the radius plot for each case tends to flatten out for larger numbers of quarks. **case1nf1xmax** refers to multi-quark families of charmonium with no degeneracy. In this case all the charm quarks have the same spin and hence cannot occupy the same state. **case1nf2xmax** is instead the plot of the charmonium family with a degeneracy of two. In this case, spin up and

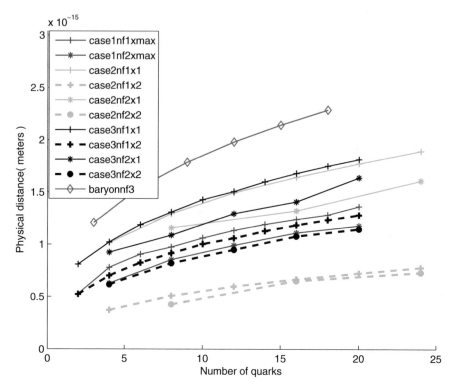

Fig. 2.9 Physical distance vs. quark content, charm and light quark–antiquarks

down is assigned to a pair of charm quarks, and thus they do not occupy the same state. The physical radius of **case1nf1xmax** being larger than **case1nf2xmax** reflects this fact. Note that the dotted lines refer to the inner boundary associated with the charmed quark in Cases 2 and 3. For Case 2 the difference between dotted and continuous lines is the largest. That means, like the Z-meson, all the higher quark family members have heavy charm–anticharm concentrated at the center, while light and antilight quarks are spread throughout. **case2nf1x2** refers to the radius plot of the inner boundary of the multi-quark family of Z-mesons with degeneracy of one, while **case2nf2x2** refers to the same plot with degeneracy of two, and similarly for Case 3. In all cases we see that the plot of physical radius is larger for degeneracy one compared to degeneracy two. The Z- and D-meson family members are found to have equally large outer boundaries. For a given quark number, the outer radius of all types of mesons was found to be smaller than the generic baryon. While comparing the physical radius between charm and bottom sectors, we noticed outer boundary is almost equal. However, inner boundary is smaller for bottom sector as can be expected from heavy quarks. Note that these radii are determined by the size of the TF functions; the electromagnetic radii have not yet been evaluated.

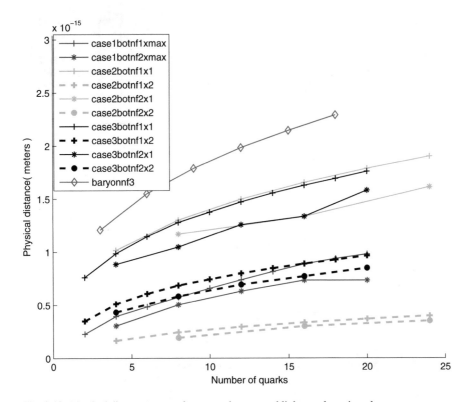

Fig. 2.10 Physical distance vs. quark content, bottom and light quark–antiquarks

There are three types of energies in this model: kinetic, potential, and volume. The kinetic energy per quark (in MeV) depends strongly on the meson family, as seen in Figs. 2.11 and 2.12. We see that the energy per quark is relatively small and tends to decrease slowly, which seems to provide some justification for this nonrelativistic model. Figures 2.13 and 2.14 show the corresponding graph for the potential energy. Figures 2.15 and 2.16 show the graph for volume energies. The graph is similar when we interchange bottom with charm quarks.

Figures 2.17 and 2.18 shows the total energy per quark without the mass term, i.e., the sum of kinetic, potential, and volume energy plotted against the quark content. The generic baryon rises slowly for increasing quark content, implying these are unstable; i.e., a higher quark content state can decay into lower members of the same family. Case 1 mesons rise quickly, and then continue the rise more slowly; these are also unstable. In contrast, the non-degenerate Case 2 mesons do not rise so quickly but are rather similar to the baryon. However, the degenerate Case 2 has an interesting pattern. We can observe negative slope from $n = 8$ to $n = 16$ when we account for degeneracy Fig. 2.19 and 2.20.

Figures 2.19 and 2.20 are our final result. Here we have plotted five points for Case 2 and six points for Case 3 one with charm-light quarks and another with

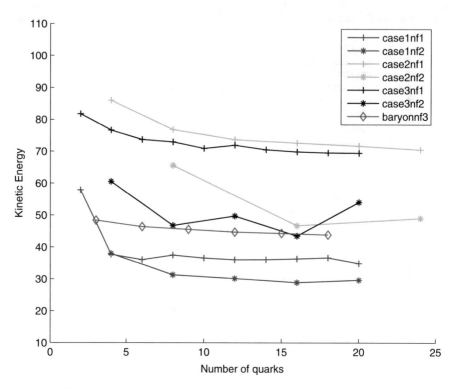

Fig. 2.11 Kinetic energy per quark vs. quark content, charm and light quark–antiquarks

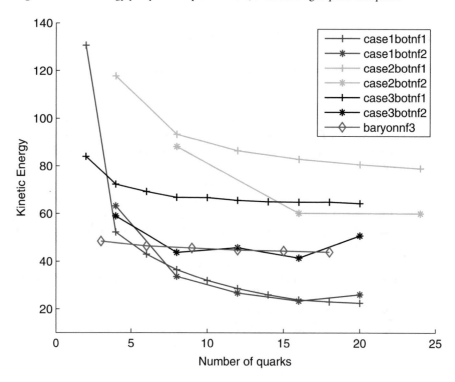

Fig. 2.12 Kinetic energy per quark vs. quark content, bottom and light quark–antiquarks

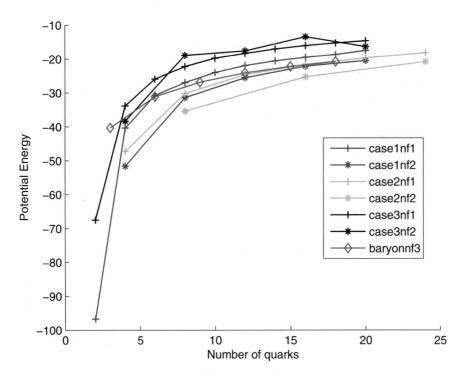

Fig. 2.13 Potential energy per quark vs. quark content for charm and light quark–antiquarks

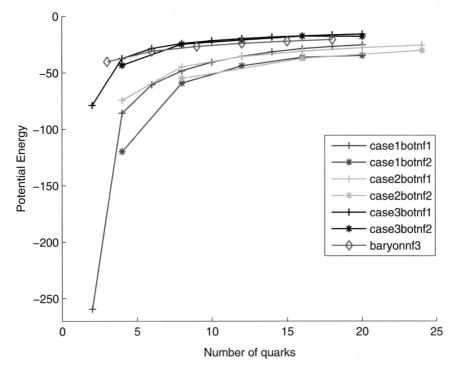

Fig. 2.14 Potential energy per quark vs. quark content for bottom and light quark–antiquarks

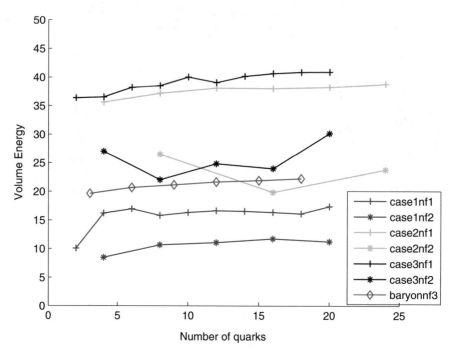

Fig. 2.15 Volume energy per quark vs. quark content for charm and light quark–antiquarks

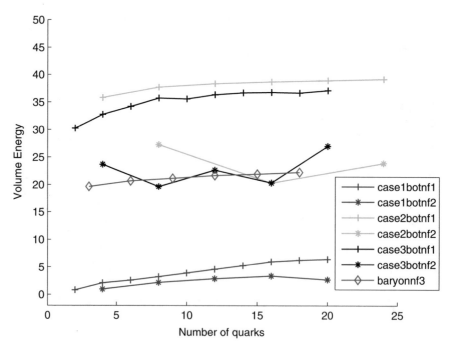

Fig. 2.16 Volume energy per quark vs. quark content for bottom and light quark–antiquarks

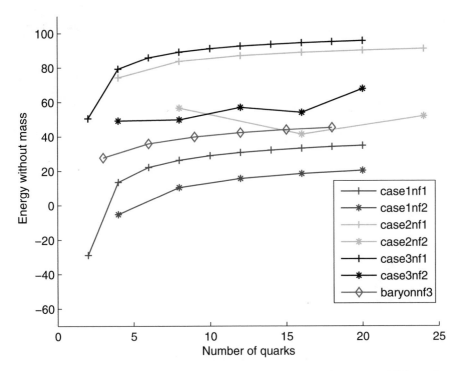

Fig. 2.17 Total energy per quark without mass term vs. quark number for charm and light quark–antiquarks

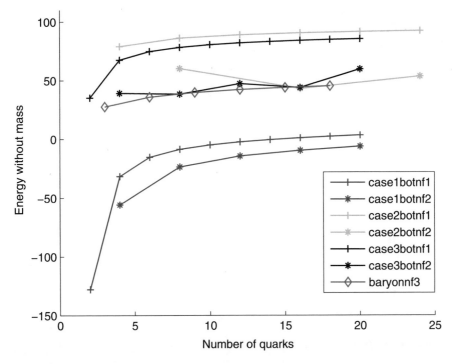

Fig. 2.18 Total energy per quark without mass term vs. quark number for bottom and light quark–antiquarks

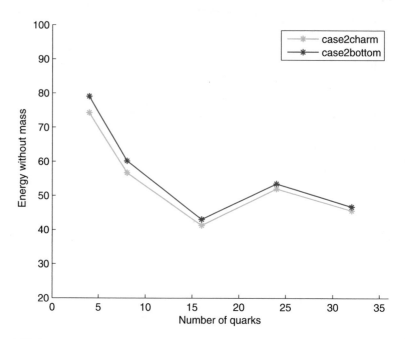

Fig. 2.19 Comparison of charm-light vs. bottom-light quarks

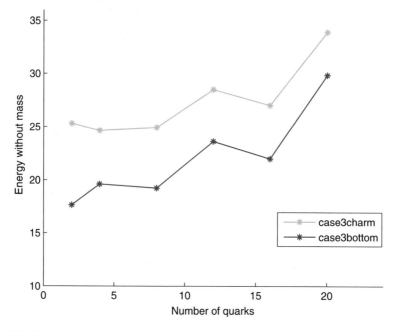

Fig. 2.20 Comparison of charm-light vs. bottom-light quarks

bottom-light quarks. For Case 2, we can see negative slope from $n = 4$ to $n = 8$ and also from $n = 8$ to $n = 16$. After that slope tends to get positive. This hints possible existence of stable multi-meson with $n = 8$ and also at $n = 16$. For Case 3 charm, we can see negative slope from $n = 2$ to $n = 4$. This hints possible existence of a tetraquark. We have assumed mass of the up and down quarks to be equal. So, at $n = 8$, we have assumed charm, anticharm, light, and antilight quarks all have degeneracy of two. While at $n = 16$ we have 2 charm and anticharm quarks both with degeneracy of two. However, we can have both light and antilight quarks with degeneracy of four. Degeneracy factor of four contribute to smaller values of binding energy per quark. Due to the limitation in our model, we were not able to investigate the intermediate states.

2.10 Conclusions

We have initiated the study of multi-quark mesons using the TF quark model. After specifying the explicit interactions and summing on colors, we have formulated system interactions and energies in a mean field approximation. We have investigated three cases of mesonic states: charmonium family, Z-meson family, and D-meson family, as well as their bottom-quark analogs. We have not yet included explicit spin interactions in our model, but we can take one level of degeneracy into account in our two TF function construction.

We have observed interesting patterns of single-quark energies. Similar to our findings for baryons, the energy per quark is slowly rising for Case 1 mesons, implying instability. The generic baryon also slowly rises. These findings are dynamical results and could not have been predicted from first principles. Ref. [1] shows convincing evidence from the lattice that the multi-D and multi-B meson tetraquarks are actually stable. Our Case 3 finding for tetraquarks cannot be considered definitive because of the lack of spin splitting terms in our interaction, but the overall trend argues against a family of such higher quark number states for the bottom sector. However, Case 3 involving charm quarks shows relative stability. Our Case 2 findings are the most interesting of our study. It is the only case where we see an actual *decrease* in the energy of introduced quark pairs, in both charm and bottom sectors. Superficially, this would indicate that stable octaquark and hexadecaquark versions of the charmed and bottom Z-meson exist. However, the energy trend is not smooth and the spin interactions have also not been taken into account.

Acknowledgements We thank the University Research Committee of Baylor University for their support of this project. We would like to thank the Texas Advanced Computing Center for account support. We also would like to thank N. Mathur for useful discussions and G. Chandra Kaphle for helpful considerations. Finally, we would like to thank Everest Institute of Science and Technology (EVIST), Kathmandu, Nepal and Neural Innovations LLC, Hewitt, Texas for assistance during research.

References

1. P. Junnarkar, M. Padmanath, N. Mathur, arXiv:1712.08400
2. P. Bicudo, J. Scheunert, M. Wagner, Phys. Rev. **D 95**, 034502 (2017)
3. A. Francis et al., Phys. Rev. Lett. **118**(2017), 142001 (2017)
4. P. Bicudo et al., Phys. Rev. **D 96**, 054510 (2017)
5. S. Prelovsek, L. Leskovec, Phys. Lett. **B 727**, 172 (2013)
6. S.K. Choi et al., (Belle), Phys. Rev. Lett. **91**, 262001 (2003)
7. S.K. Choi et al., (Belle), Phys. Rev. Lett. **100**, 142001 (2008)
8. A. Bondar et al., (Belle), Phys. Rev. Lett. **108**, 122001 (2012)
9. Z.Q. Liu et al., (Belle), Phys. Rev. Lett. **110**, 252002 (2013)
10. M. Ablikim et al., (BESIII), Phys. Rev. Lett. **110**, 252001 (2013)
11. M. Ablikim et al., (BESIII), Phys. Rev. Lett. **115**, 112003 (2015)
12. R. Aaij et al., (LHCb), Eur. Phys. J. **C72**, 1972 (2012)
13. R. Aaij et al., (LHCb), Phys. Rev. Lett. **110**, 222001 (2013)
14. R. Aaij et al., (LHCb), Phys. Rev. Lett. **112**, 222002 (2014)
15. R. Aaij et al., (LHCb), Phys. Rev. Lett. **115**, 072001 (2015)
16. R. Aaij et al., (LHCb), Phys. Rev. Lett. **118**, 022003 (2017)
17. E. Farhi, R. Jaffe, Phys. Rev. **D 30**, 2379 (1984)
18. J. Madsen, J.M. Larsen, Phys. Rev. Lett. **90**, 121102 (2003)
19. J. Madsen, Phys. Rev. Lett. **100**, 151102 (2008)
20. O. Kiriyama, Phys. Rev. **D 72**, 054009 (2005)
21. W. Wilcox, Nucl. Phys. **A 826**, 49 (2009)
22. S. Baral, W. Wilcox, Nucl. Phys. **A 982**, 731–734 (2019)
23. S. Baral, https://hdl.handle.net/2104/10503
24. Q. Liu, W. Wilcox, Ann. Phys. **341**, 164 (2014).Erratum to be published
25. P. Sikivie, N. Weiss, Phys. Rev. **D 18**, 3809 (1978)
26. S. Baral, W. Wilcox (DPF 2017) arXiv:1710.00418
27. W. Wilcox, S. Baral (Schwinger Centennial Conference) arXiv:1807.08626
28. M. Tanabashi et al., (Particle Data Group), Phys. Rev. D **98**, 030001 (2018), Sec. 90, Tables 90.2 and 90.3
29. L. Maiani et al., Phys. Rev. **D 89**, 114010 (2018)

Printed in the United States
By Bookmasters